Wasps

Published in the United States and Canada
in 2021 by Princeton University Press
41 William Street
Princeton, New Jersey 08540
press.princeton.edu

First published 2021 by UniPress Books

Published by PUP by arrangement with
UniPress Books Ltd.
www.unipressbooks.com

Commissioning Editor: Kate Shanahan
Project Manager: Natalia Price-Cabrera
Design & Art Direction: Sandra Zellmer
Illustration: Sandra Pond

Library of Congress Control Number
2020942974
Hardcover ISBN 978-0-691-21142-8
Ebook ISBN 978-0-691-21864-9

Printed in China

10 9 8 7 6 5 4 3 2 1

Wasps

The Astonishing Diversity of a Misunderstood Insect

ERIC R. EATON

PRINCETON UNIVERSITY PRESS
PRINCETON AND OXFORD

Contents

Introduction

Those who have nobly suffered the stings and harrowing attacks of yellowjackets or hornets may have a low opinion of wasps, but it is the mission of this book to turn fear and loathing into fascination and admiration. It may come as a surprise to learn that the overwhelming majority of wasp species lead solitary lives rather than dwelling in paper palaces with queens and workers; or that not all wasps can sting. Even for those species that do, it is only females that possess a stinger and venom glands.

Bees and ants are close relatives of wasps, yet we vilify wasps, worship bees, and proclaim ants to be the pinnacle of insect evolution. Entomologists have done a rather poor job of educating the public about the merits of wasps, but the media is of little help, either. Complicating matters are invasive species that, once established on foreign soil, wreak disproportionate havoc in their new homeland. Couple that with the ability of some wasps to exploit weaknesses in the behavior and immune system of people, and you have a recipe for a dim if not hostile human viewpoint.

Meanwhile, reality is vastly different. Wasps, not just bees, are invaluable pollinators of plants. Minute, parasitic wasps barely visible to the human eye are propagated in laboratories for dispersal into crops where they control pests with more precision than chemical applications could ever accomplish, and with near zero side effects. Hardly any other insect exists that does not have at least one wasp enemy, and the specific nature of most host-parasite relationships makes wasps an indispensable part of our pest-control arsenal.

Even venom, the one product of wasps that is the centerpiece of our hatred for them, can be an asset to medicine and other scientific disciplines. We have barely begun to analyze and make use of these amazing compounds and already the possibilities are highly promising.

So complex are the behaviors of even solitary wasps, and so plastic are their supposedly robotic neurological programs, that we are forced to redefine instinct and intelligence. The study of beewolf wasps led to formal recognition of ethology, the study of animal behavior, and earned a share of a Nobel Prize for Niko Tinbergen. Alfred Kinsey, the scholar of human sexuality, got his start studying gall wasps.

Periodic personal experiences with wasps may be painful, but our civilization would not be advancing without these insects, and natural ecosystems would collapse were it not for the diversity of wasp species we see today. It is in our best interest to devote more energy to their study, and open our hearts and minds to their positive attributes.

Pollinator
A female sand wasp, *Bembix* sp., drinks nectar from a flower to fuel her flight, and accomplishes pollination in the process.

Entomologists, the scientists who study insects, define "wasp" more broadly than the rest of us. We assume the term to mean social wasps like yellowjackets, hornets, and paper wasps. This accidental tunnel vision excludes an astonishing array of other wasps more colorful, larger, or smaller, and several orders of magnitude more fascinating and beneficial than the few species that plague picnics or barbecues. We cower when confronted by the smallest of spiders, but a female tarantula hawk wasp hunts down the largest of arachnids. We cannot recall where we parked our vehicle, but a sand wasp can locate her hidden burrow in a dune. Wasps own sophisticated senses we can only dream of. Most can fly. There is much to admire, if not envy.

Each chapter in this book explores attributes of wasps that set them apart from all other animals. Their anatomies and life cycles share similarities with other insects, but the diversity and efficiency of their lifestyles and behaviors are unsurpassed. Wasps have influenced the evolution of flowers, caused unrelated insects to evolve a nearly identical resemblance to them, and exerted pressure on host and prey insects to evolve new survival strategies. Wasps weaponized their egg-laying organs. Some species have evolved symbiotic relationships with fungi, or viruses, and bent them to their advantage. Many species are allies of man in pest control, medicine, and invention. Wasps are the epitome of the warrior spirit, female empowerment, and relentless dedication to survival. Wasps excel as opportunists and exploiters.

Wasps have achieved their success through involuntary adaptations over time, evolving subtle changes to their physical bodies and genetic heritage. We, on the other hand, are capable of rapid social and economic adaptations by executing conscious choice in our habits as producers and consumers. One could argue that our male-dominated societies are more ruthless and destructive than any matriarchal "murder hornet" colony. This is a crucial period in the history of humanity and the natural history of the planet. Will we choose wasps to be a part of it?

Scientists argue that despite anecdotal evidence from Germany, Australia, Canada, Puerto Rico, and a few other locations, there is not quantitative nor qualitative data to support the assertion of an "insect apocalypse." There is, however, widespread agreement that we know enough to act now to prevent further erosion of insect abundance and diversity. What we cannot live without is obvious. Our psyche suffers when deprived of biodiversity, and our physical lives become threatened when ecosystem services like pollination, natural pest control, seed dispersal, and organic decomposition are compromised. There is no alternative but to preserve and protect all remaining extant species, wasps included.

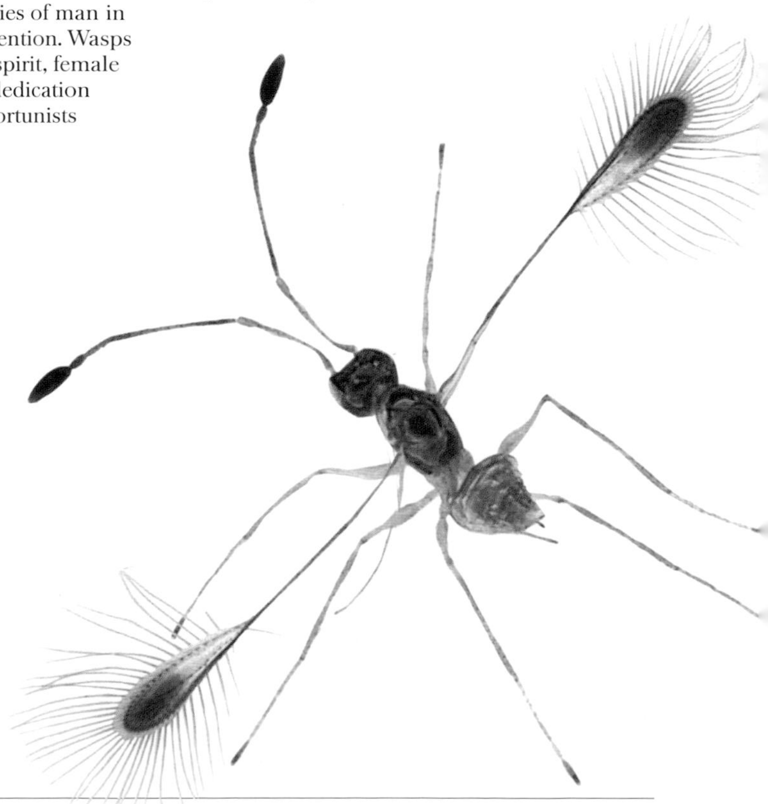

Larger Than Life
The face of a cuckoo wasp (left), *Stilbum* sp., is dominated by its compound eyes and antennae. Wasps possess unexpected beauty, grace, and durability

Fairyfly
Mymar pulchellum, a nearly microscopic wasp found in Europe, Asia, and North America (right), is a parasitoid of the eggs of leafhoppers and delphacid planthoppers.

1

Evolution
The Origin of Wasps

How Wasps Evolved

Evolution is the diversification of organisms through heritable traits. It is driven by opportunities to occupy newly opened niches in ecosystems, competition for those niches, the need to avoid mortality from predators, parasites, and diseases. The need to adapt to changes in abiotic factors such as composition of the atmosphere, and also genetic mutations, genetic recombination, and related changes in genes are additional drivers of evolution. It is a non-stop, often non-linear phenomenon that still occurs, albeit at a rate that is barely discernible, especially in the digital Anthropocene epoch of instantaneous results.

Cladogram
This simplified cladogram demonstrates one current interpretation of the relationships between various categories of Hymenoptera. Each node represents a hypothetical ancestor to the lineages branching from it. Our scientific understanding of these relationships changes constantly. *Anthophila is a clade under the superfamily Apoidea.

Insects are abundant in the fossil record, perhaps surprising given our assumption of their fragility. But fossils are only one tool in our approach to understanding the evolution of wasps. Our comprehension is also informed by observations of external physiology (morphology) of living and extinct specimens, and by analysis of nuclear, ribosomal, and/or mitochondrial DNA in living species that reveal previously unknown relationships between higher levels of classification over time. It is a dynamic scientific endeavor, constantly revised in the face of new evidence. Competing hypotheses are the order of the day, even when more than one method of evaluation is applied.

The "start here" sign for wasps on the geological timeline is, according to fossils, probably somewhere in the late Permian period, 260–270 million years ago. They perhaps originated from the now-extinct Parasialidae, ancestors of today's alderflies.

There is widespread agreement that the first recognizable wasps emerged in the middle of the Triassic period of the Mesozoic era, about 235 million years ago, as indicated by the fossil record. These were sawflies in the superfamily Xyeloidea. Molecular DNA research contradicts this, however, suggesting xyelids may have an earlier origin in the late Carboniferous period, roughly 311 million years ago. Yet another alternative theory proposes the xyelids stemmed from an ancestor related to scorpionflies (Mecopteroidea). Indeed, sawfly larvae and scorpionfly larvae greatly resemble moth or butterfly caterpillars in their nearly identical external anatomy and largely herbivorous feeding habits. The precise relationship of xyelids to the remaining sawfly superfamilies, Tenthredinoidea and Pamphiloidea, is unclear. What we call wasps today—the stinging parasitoids, plus ants and bees—began diverging around the beginning of the Jurassic period, slightly less than 200 million years ago, we think.

Wasps, ants, and bees are placed in the order Hymenoptera, and recognized as the first major branch of all insects that undergo complete metamorphosis (egg, larva, pupa, adult). The order name is Greek and translates to "marriage wing," a reference to how the forewing and hindwing are joined to produce one functional flying pair.

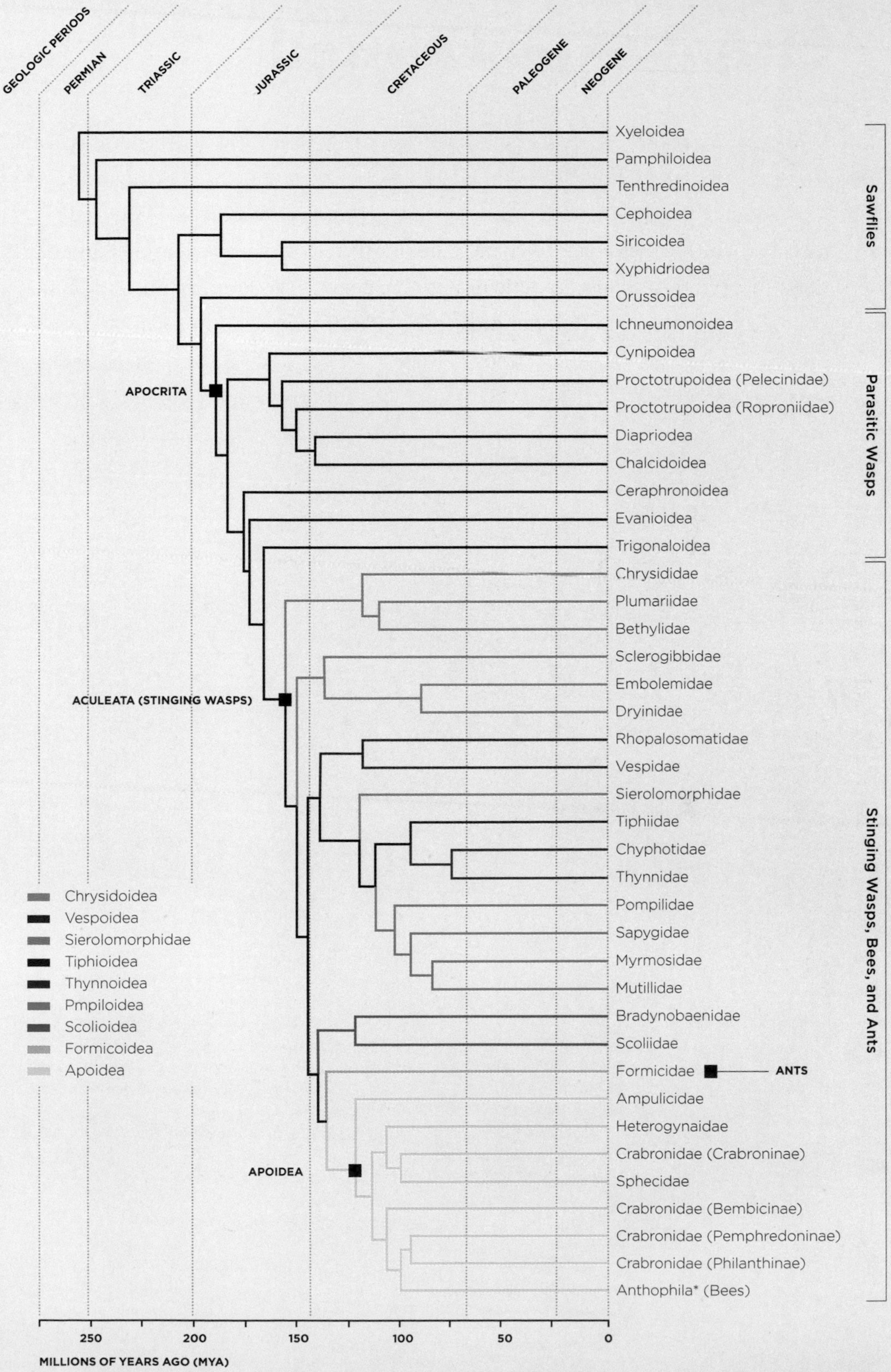

GEOLOGIC PERIODS
PERMIAN
TRIASSIC
JURASSIC
CRETACEOUS
PALEOGENE
NEOGENE
Xyeloidea
Pamphiloidea
Tenthredinoidea
Cephoidea
Siricoidea
Xyphidriodea
Orussoidea
Ichneumonoidea
Cynipoidea
Proctotrupoidea (Pelecinidae)
Proctotrupoidea (Roproniidae)
Diapriodea
Chalcidoidea
Ceraphronoidea
Evanioidea
Trigonaloidea
Chrysididae
Plumariidae
Bethylidae
Sclerogibbidae
Embolemidae
Dryinidae
Rhopalosomatidae
Vespidae
Sierolomorphidae
Tiphiidae
Chyphotidae
Thynnidae
Pompilidae
Sapygidae
Myrmosidae
Mutillidae
Bradynobaenidae
Scoliidae
Formicidae
ANTS
Ampulicidae
Heterogynaidae
Crabronidae (Crabroninae)
Sphecidae
Crabronidae (Bembicinae)
Crabronidae (Pemphredoninae)
Crabronidae (Philanthinae)
Anthophila* (Bees)
Sawflies
Parasitic Wasps
Stinging Wasps, Bees, and Ants
APOCRITA
ACULEATA (STINGING WASPS)
APOIDEA
Chrysidoidea
Vespoidea
Sierolomorphidae
Tiphioidea
Thynnoidea
Pmpiloidea
Scolioidea
Formicoidea
Apoidea
250
200
150
100
50
0
MILLIONS OF YEARS AGO (MYA)

Trapped in Amber

When most people think of insect fossils, they think of those inside amber. From the movie *Jurassic Park* to articles in newspapers, and popular nature and science magazines, insects in amber captivate the imagination and make for jaw-dropping graphics. Amber is fossilized plant resin. Insects trapped in amber are termed "inclusions" because the actual creature, or parts thereof, are encased in the resin. Amber deposits range in timespan from the Carboniferous period (359 million years ago to 299 million years ago) to the Holocene, our current geological epoch, from roughly 11,650 years ago to the present day. The largest and most well-known amber deposit is from the southern coast of the Baltic region of Europe, and has its origins in the sap of coniferous trees, probably pines. Dominican amber represents fossilized resin from broad-leaved, legume trees. Amber deposits also occur in Austria, England, Lebanon, Jordan, and Japan.

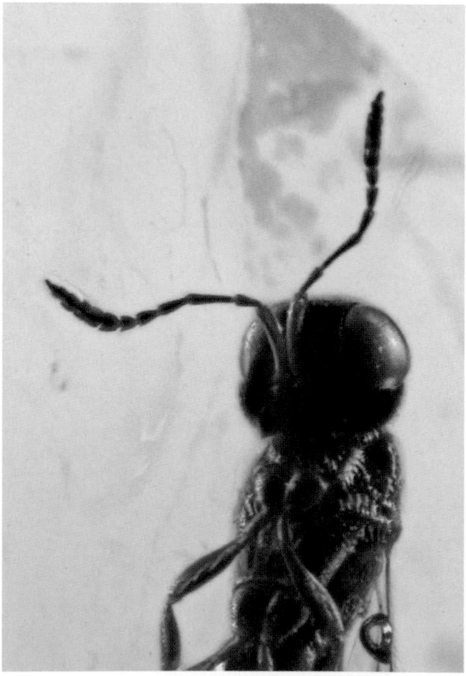

Amber fossils are particularly valuable because of their three-dimensional quality, allowing scientists to view specimens from several angles and discern details of antennae, setae, and other delicate parts of insect anatomy that are rarely preserved in stone fossils. However, gas bubbles and debris frequently obscure key parts of insect fossils in amber.

Amber is not always that color. It can be various shades of gold or butterscotch, or even green or blue. It is sold commercially at gem shows, online, and through other retail outlets. Some is fashioned into earrings, brooches, or other accessories, so the next major discovery in the realm of fossil insects may be residing in your jewelry box or dresser drawer.

Ancient Species
A wasp of the family Platygastridae (subfamily Scelioninae), preserved in Dominican amber. The clarity and detail are typical of amber fossils.

Entombed
A tiny wasp, possibly of the family Diapriidae or a precursor to that family, trapped in Baltic amber.

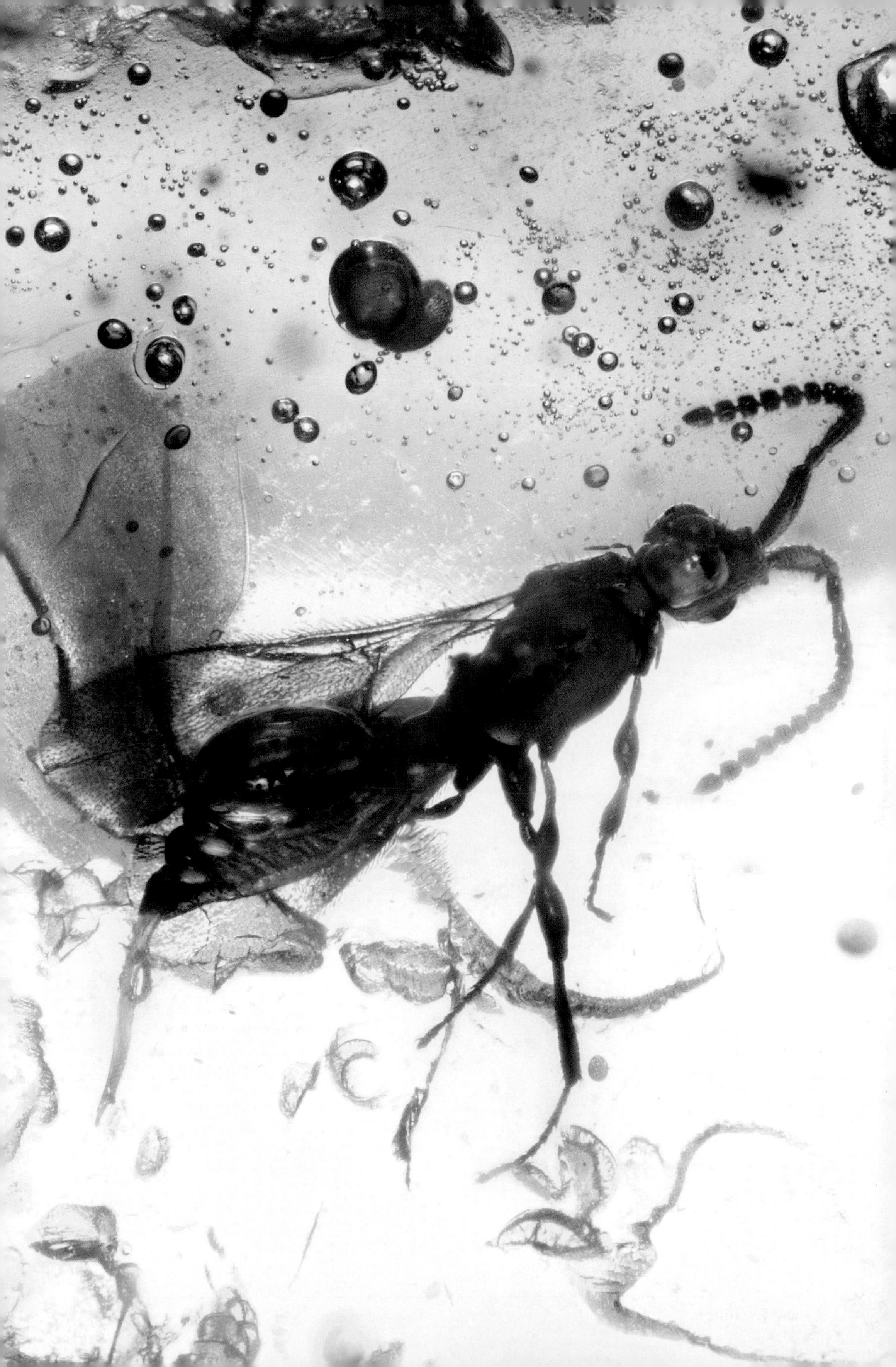

Fossils in Stone

Amber may get all the glory when it comes to insect fossils, but there are fossils of insects in rocks as well. Deposits of these occur across the globe and represent stretches of time from the Carboniferous period (359 million years ago to 299 million years ago) to recent times (less than 10,000 years ago). Stone fossils of insects take two forms: impressions and compressions. Both are usually the result of sediments quickly covering an organism, from either flooding, or falling volcanic ash.

Fossil Hornet
This specimen of *Palaeovespa florissantia* (left), an ancestor of modern day yellowjackets and hornets, was discovered in Florissant, Colorado, U.S.A.

Wasp in Rock
This specimen of *Archisphex* (facing, top right) from Nova Olinda in Brazil is an example of a compression, like the hornet fossil to the left.

Eocene Mystery
An unidentified wasp (facing, bottom right) of the Eocene epoch is preserved as a roughly 50 million years old fossil from the Green River Formation in Wyoming, U.S.A.

Impressions occur when an insect leaves an imprint on a surface; the insect disintegrates or washes away, but leaves a shadow of its former self. Imagine placing your hand in wet concrete and then departing. The cement hardens, and although you are now absent, the evidence of your visit remains.

Compressions include the actual insect itself —the fossilized exoskeleton. Even fine details in the cuticle may still be evident. One of the most spectacular reservoirs of insect fossils in stone is in Florissant, Colorado, USA. The iconic fossil from this deposit, and insignia for the Florissant Fossil Beds National Monument, is none other than a social wasp, *Palaeovespa florissantia*. This particular fossil dates to the Eocene epoch, somewhere between 44 and 34 million years ago in the Paleogene period of the Cenozoic era.

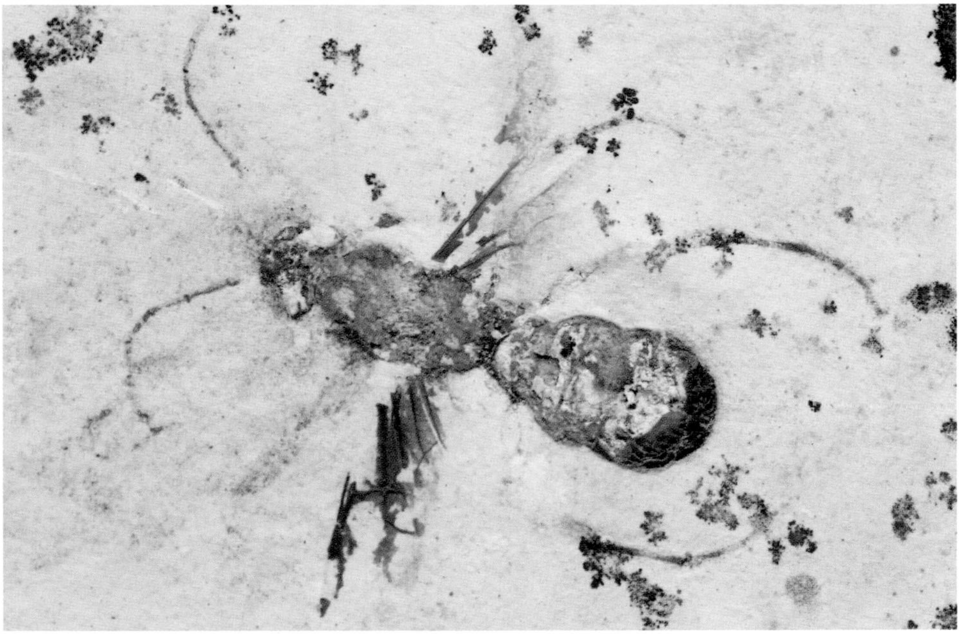

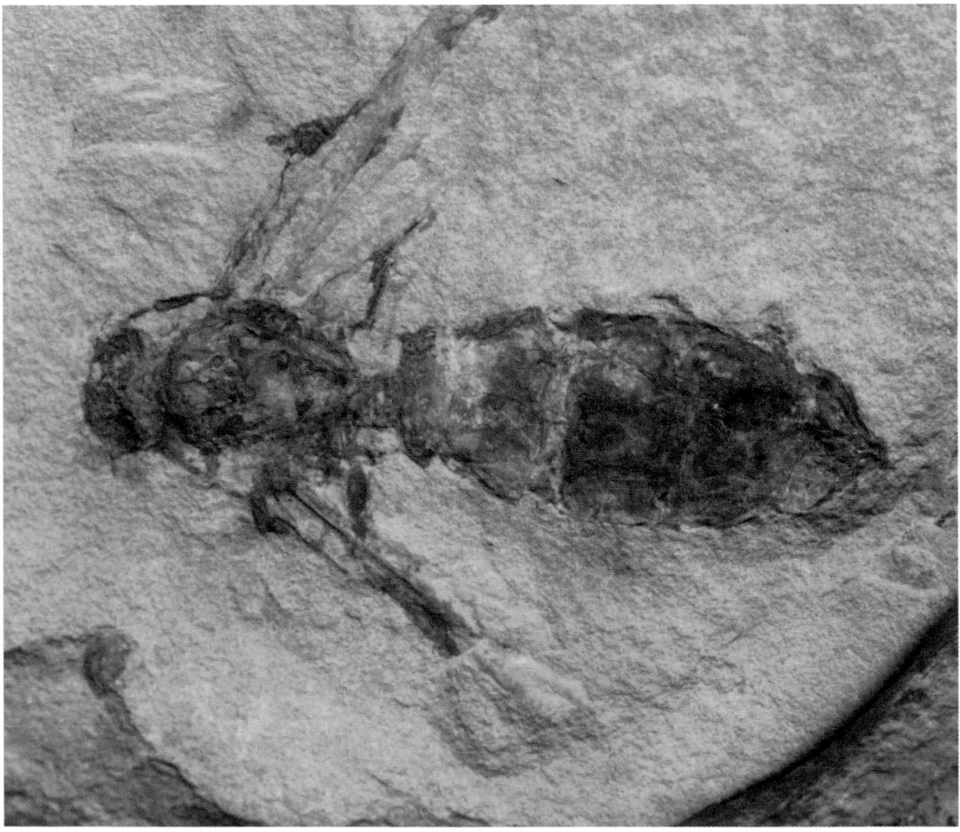

Are Bees Just Hairy Wasps? Yes!

Sawflies evolved into parasitoid wasps, which then gave rise to the stinging parasitoids we commonly refer to as "true wasps." From those stinging wasps arose bees, beginning sometime in the middle of the Cretaceous period, the last period in the Mesozoic era. That translates to approximately 125 million years ago.

Cuckoo Bee
A nomad bee, *Nomada* sp. (below) looks much like a wasp in color, shape, and lack of dense hair.

Honey Bee
Native to the Old World, honey bees (*Apis* spp.) are social wasps that we value for their pollination of crops and production of honey. Shown here is the "pollen basket," found on the hind leg, where pollen and nectar are collected (left).

Tongue Out
A bumble bee (*Bombus* sp.) displays the highly modified mouthparts it uses to extract nectar from flowers. Unlike bees, Most wasps lack a specialized "tongue."

The driving force behind the evolution of bees was the evolution of flowers, and vice versa. Floral diversity began to literally blossom when wasps and other insects started feeding on nectar inside simple, cryptic flowers, accomplishing pollination as a by-product of their visits. Flowers reinforced this behavior by evolving to produce more nectar, and becoming more colorful and fragrant. Some wasps then switched from hunting prey for their offspring to harvesting protein-rich pollen instead. The fossil record for early bees is largely blank because flowers evolved in arid landscapes that lacked conditions conducive to the fossilization process.

Look closely at bees and you will see they are similar to wasps in anatomy and behavior, with a few key differences. Individual hairs on bees are branched, the better to trap pollen, while the hairs of wasps are unbranched. Naturally there are hairless or nearly hairless wasps, and many bee species are also sparsely hairy. Certain segments in the mouthparts of bees are modified into a "tongue" to reach nectar reservoirs inside flowers. Some wasps have similar configurations, but bees generally have more complex, longer versions.

The bodies of bees are usually modified to carry pollen loads. Solitary bees have a dense brush of hairs, called a scopa, for collecting dry pollen. The scopa is located on either the hind legs (most solitary bee families), or on the underside of the abdomen (leafcutter bees, mason bees, resin bees, and their kin). Social bees typically have the tibia and basitarsus (first segment of the tarsus) of the hind leg flattened, expanded, and fringed with long hairs to form a "pollen basket" into which a sticky paste of pollen and nectar is collected. Bees and wasps that lack pollen-trapping hairs may ingest pollen and regurgitate it when provisioning their nests.

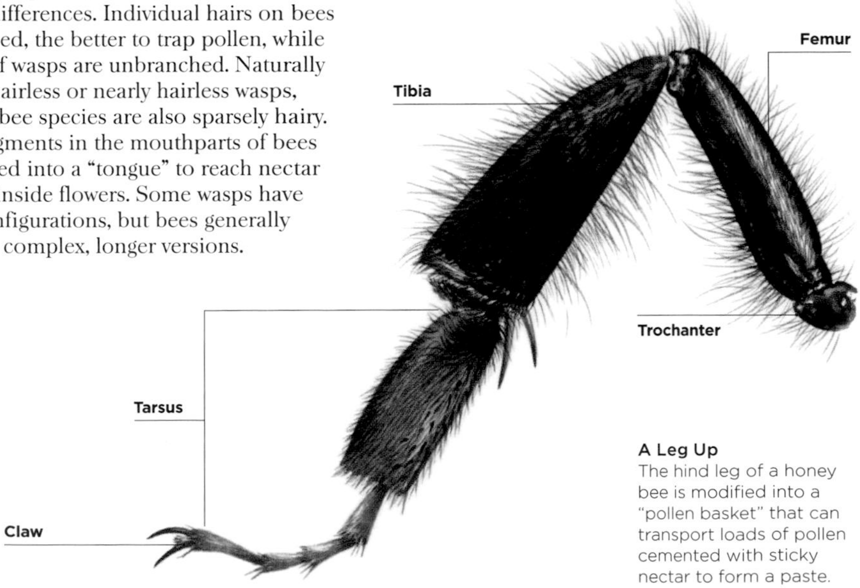

A Leg Up
The hind leg of a honey bee is modified into a "pollen basket" that can transport loads of pollen cemented with sticky nectar to form a paste.

Advent of the Social Lifestyle

Sociality ranges from subsocial behavior, like lengthy care of offspring, to communal nesting where several females tend their own offspring within a common nest. The height of social behavior is termed eusocial, and its overriding characteristic is the overlap of generations. Wasps likely exhibited their first social tendencies in the mid-Cretaceous period, roughly 115–120 million years ago. A fossilized paper wasp nest from the late Cretaceous lends credence to that. Wasps, bees, and ants have evolved some degree of social behavior over 20 separate times, but it didn't always stick. Sometimes evolutionary descendants revert to solitary lifestyles.

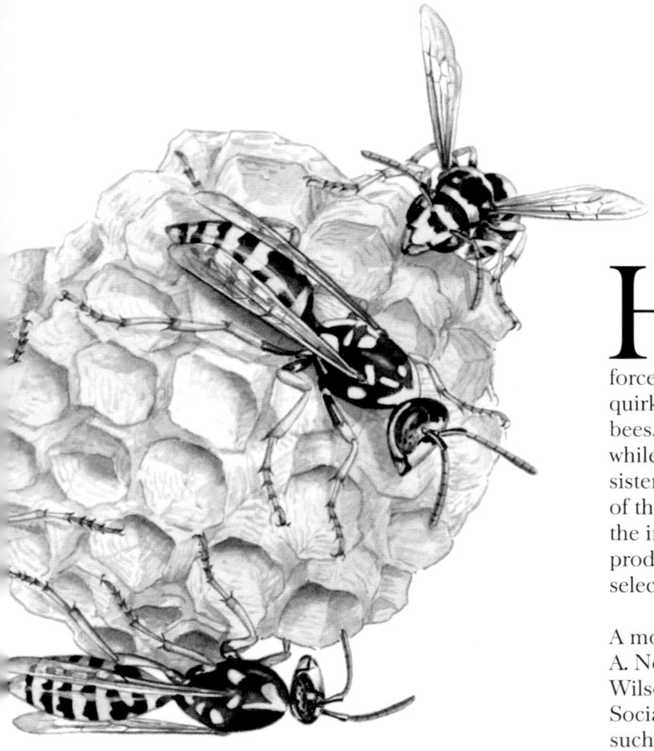

Paper Wasps
Wasps in the genus *Polistes* represent the midway point on the spectrum of sociality, with a dominance hierarchy but no dedicated queen. Nests are uncovered paper combs.

Historically, the "inclusive fitness theory" focused on how natural selection favors only closely related individuals as the driving force in the evolution of sociality. Through a quirk in sex-determination in wasps, ants, and bees, females have two sets of chromosomes while males have only one. The result is that sisters share 75 percent of their genes instead of the expected 50 percent. Insect colonies, the inclusive fitness theory asserts, are a product of that relatedness because natural selection operates by kin selection.

A more recent theory, published by Martin A. Nowak, Corina E. Tarnita, and Edward O. Wilson recognizes five steps to eusociality. Sociality likely evolved from circumstances such as scarce food resources that would tend to clump individuals of a species within a habitat. The second step would be construction of a defensible nest to guard collected food. Organisms with complex solitary behavior, such as hunting wasps with complicated maternal conduct, are likely to have a greater predisposition toward sociality than species with simpler behaviors, such as sawflies.

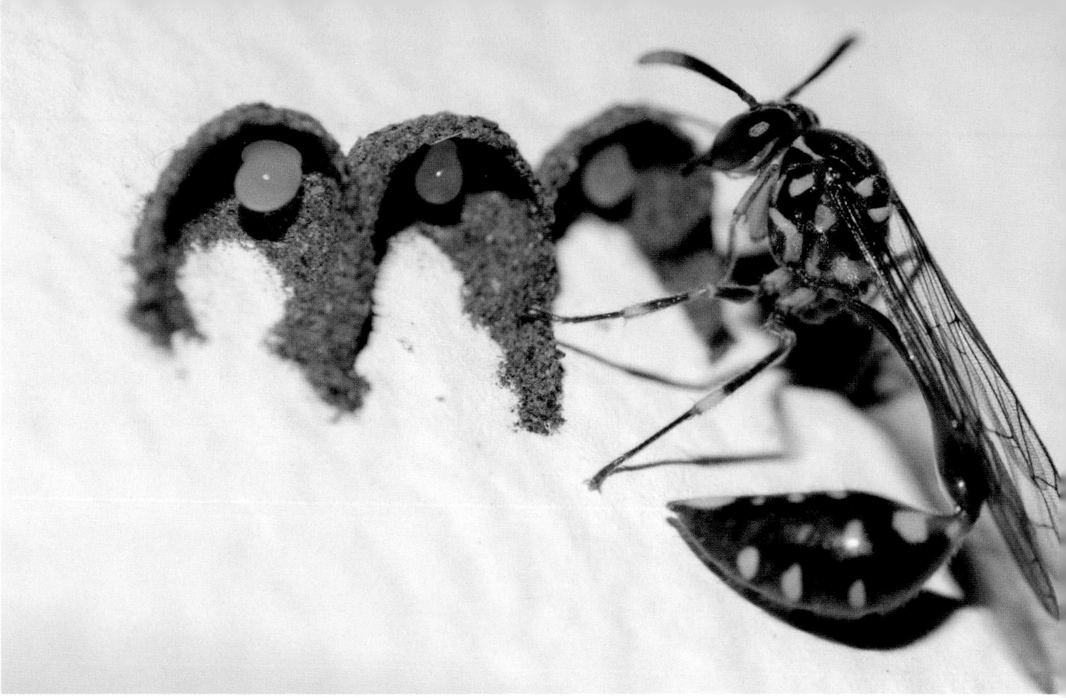

Primitively Social
The incremental success of sociality in wasps is evident in the varying levels of sophistication we see today in the family Vespidae. Most diverse in tropical regions, they range from solitary mason wasps in the subfamily Eumeninae, to primitively social Stenogastrinae, and ultimately to the enormous colonies of hornets and yellowjackets in the subfamily Vespinae. Hover wasps in the subfamily Stenogastrinae, like this *Liostenogaster* sp. from Bali, Indonesia, are barely social. Females build small groups of mud cells and show little reproductive hierarchy.

After environmentally induced groupings of individuals, leading to building a defensible nest, the third step to eusociality is likely the genetic expression of sociality through a new allele that either inhibits solitary behavior or actively predisposes social behavior. This could result from a mutation or recombination of genes, and be activated and passed to future generations only if ecological and environmental circumstances reward social tendencies. The fourth step is the influence of natural selection on the emergence of traits resulting from social interactions such as group foraging. The final step in attaining eusociality is what amounts to a feedback loop: gene-encoded flexibility for social behavior results in division of labor among colony members, for example, which becomes reinforced in the genetic blueprints for future generations. This reaches its zenith in ant species that exhibit polymorphism with enormous "soldier" workers, foraging workers of medium size, and some small individuals that work inside the nest.

The new theory also supports the more long-standing notion that an insect colony is itself a "superorganism" upon which natural selection operates.

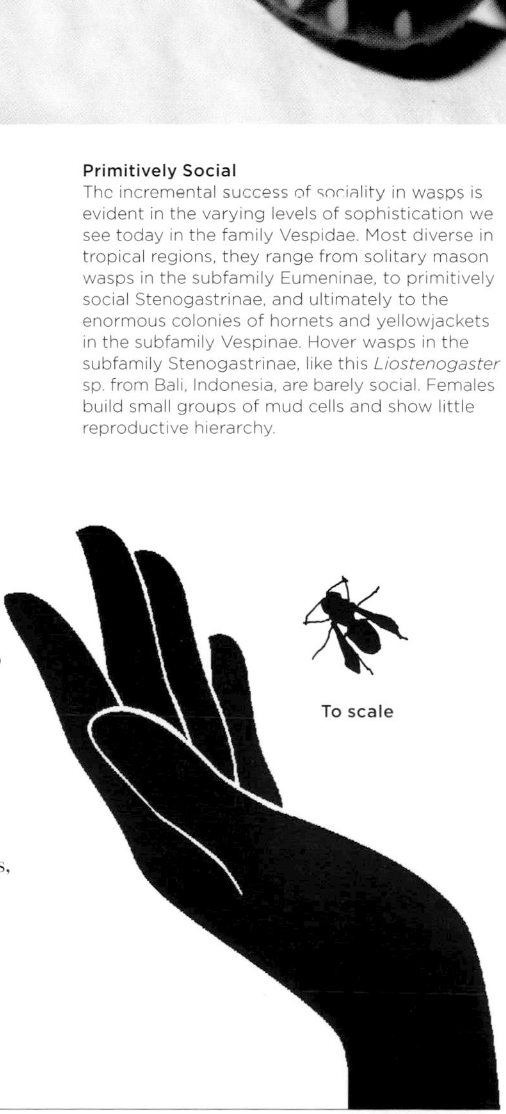

Classification of Wasps

The science of classification is called taxonomy. A taxon can be any level of classification, from kingdom to species. The *way* we classify organisms is influenced by our interpretation of the evolutionary paths those organisms took to arrive at their existence in time. These relationships can be illustrated by phylogenetic trees, or cladograms. Before molecular DNA research entered the picture, all graphic representation was in the form of evolutionary trees, the length of each branch reflecting no significance at all, elapsed time, or the degree to which a given character or characters had changed from point of origin to the next node or terminal taxon.

A cladogram depicts an arrangement of clades, each clade representing a single ancestor taxon and all of its descendant taxa. One clade is the subset of a larger clade, and so on. While some scientists consider cladograms and phylogenetic trees as synonymous, others define cladograms as representing hypothetical evolutionary relationships and phylogenetic trees as established evolutionary pathways.

Why all the Latin? The founding father of taxonomy, Swedish biologist Carl Linnaeus, chose Latin because it was already a common language among scientists of his time, minimizing confusion. Perhaps it was also out of respect for Aristotle who established some categories in his *The History of Animals*, published in 350 B.C.E. Consequently, Greek also figures prominently in scientific names. Linnaeus's landmark *Systema Naturae*, published in 1735, debuted the official system of classification that remains the standard today. It is an organizational tool that also lends itself to change as our understanding of evolutionary relationships evolves from new fossils and new insights provided by molecular DNA research.

Wasps are arranged from most "primitive" to most "advanced," but those terms are at best suggestive and at worst irrelevant. We can, however, clearly see how stinging wasps represent a drastic change from their evolutionary predecessors, and how bees sprang from the stinging wasps. Likewise, ants clearly evolved into wingless versions of yellowjackets, hornets, and paper wasps. The fact that all of these wasps exist concurrently today show that both "primitive" and "advanced" can be equally successful.

New species, genera, families, and even higher levels of classification are discovered with regularity. Many more are known to science, but remain undescribed. That is, we have not yet assigned them names, nor formally interpreted the characters that set them apart from close relatives and thus validate them as new species. There is so much reverence for the Linnaean system that the International Commission on Zoological Nomenclature sets rules for naming animals through the publishing of the International Code on Zoological Nomenclature. It also serves as a jury to resolve problems of taxonomy.

Inverse Pyramid
Each level is a subset of the category above. Placement of a species is subject to change, as are genera and families. Subfamily, and its subset "tribe," can also be referred to in wasp classification.

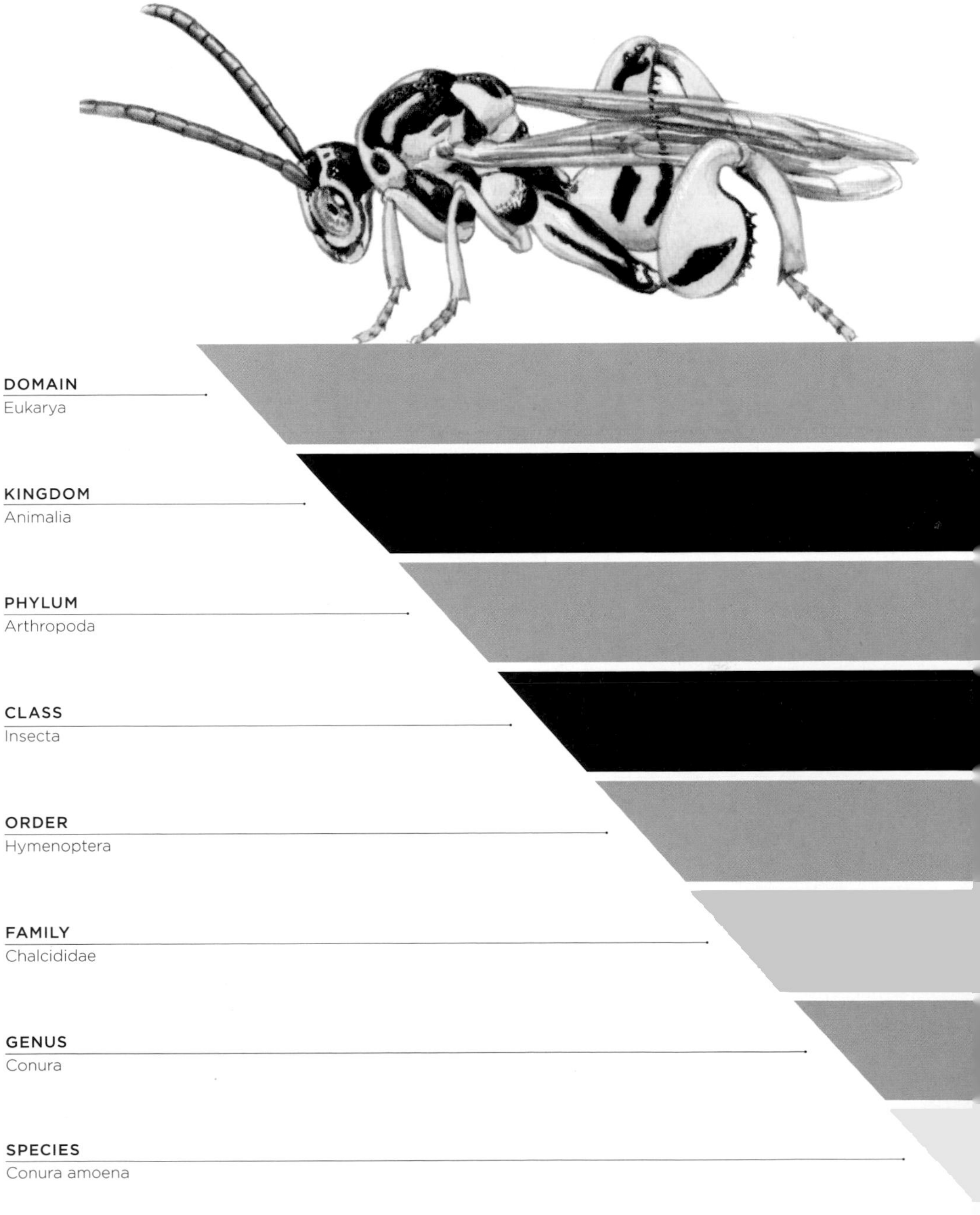

2

Anatomy
Structure and Function

How Wasps Are Built

We call other people "spineless" as an insult, but invertebrates—those animals without backbones—have done quite well. The most successful invertebrates are the arthropods, of which insects are the largest subset. Wasps have arguably elevated the insect model to its most efficient and aesthetic designs, both outside and inside.

The phylum Arthropoda is vast, encompassing insects, arachnids, crustaceans and more. Centipedes and millipedes are also arthropods, as are non-insect hexapods (springtails, symphylans, and their kin), several more obscure invertebrates, and the extinct trilobites. The word arthropod translates as "jointed foot," and all arthropods share the character of articulated appendages. Arthropods are packaged in an armor-like exterior called an exoskeleton. This is true even in the soft, squishy larva and pupa stages of insects that undergo complete metamorphosis. Arthropods also exhibit bilateral symmetry, meaning that the right and left halves of the animal mirror each other.

Insects are segregated from most other arthropods by having three distinct body sections (head, thorax, and abdomen), three pairs of legs, plus one pair of antennae. Insects are the only flying invertebrates, though not all insects have wings, including many wasps. The insect head operates most sensory organs used by the creature to navigate its environment and communicate with other members of its kind. The thorax functions as the locomotion center, to which all three pairs of legs, and wings (if any), are attached. Bringing up the rear is the abdomen, which contains most vital organs, including the excretory and reproductive systems.

The internal functioning of insects differs radically from vertebrate animals. Insects have an open circulatory system in which the blood (hemolymph) bathes every organ in the body cavity. Blood cells do not carry oxygen, either. Insects breathe through small openings called spiracles, located in the cuticle of the thorax and abdomen. Inside, a vast network of tracheae carry oxygen to every cell. Gas exchange happens by diffusion at the cellular level of delivery. This aspect of insect anatomy is one factor that limits the size of insects. We do not have wasps the size of vehicles because the tracheal system could not carry oxygen over that large an area.

The stinger is the one anatomical feature we think of immediately when we hear "wasp," but only females of some species have this weapon. Males do not. Ever. Meanwhile, there can be other graphic differences between the sexes, termed "sexual dimorphism." Various glands that produce venom, saliva, and chemicals for communication (pheromones), and modifications of legs and other appendages likewise make wasps the ultimate survival machines.

Body Plans
Wasps like this yellowjacket (top) and mud dauber (bottom) are built for durability, flexibility, and sensory processing. Each body section has a role to play, and appendages are often specialized.

Scutum
Gena
Mandible
Tergum
Sternum
Gaster

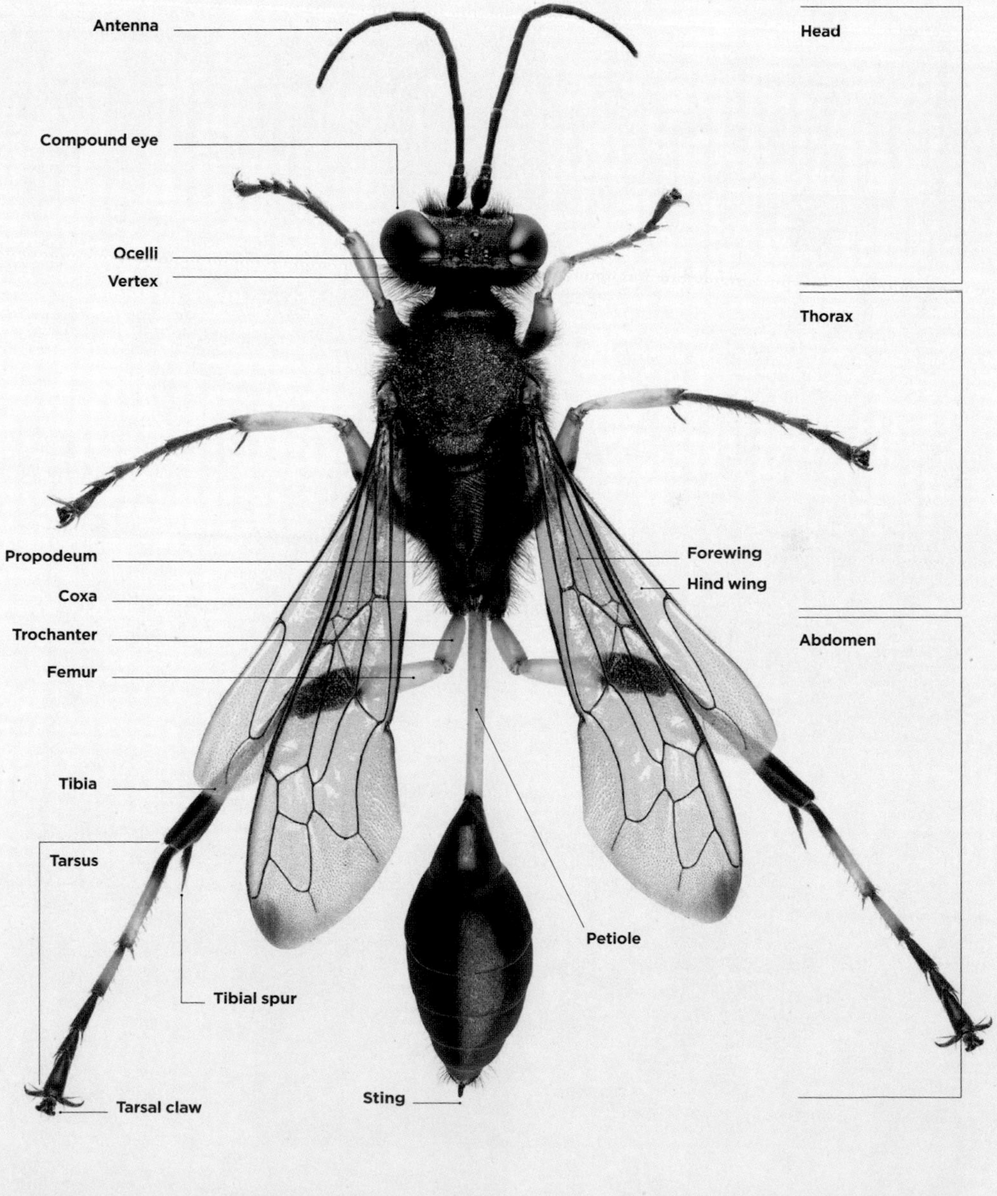
Antenna
Head
Compound eye
Ocelli
Vertex
Thorax
Propodeum
Forewing
Hind wing
Coxa
Trochanter
Abdomen
Femur
Tibia
Tarsus
Petiole
Tibial spur
Tarsal claw
Sting

Head: Sensory Input

Look a wasp in the face and several features jump out at you. Compound eyes wrap around the sides to give a nearly 360-degree view. The eye is composed of individual units called ommatidia, each of which consists of the lens facet we see on the exterior of the eye. A crystalline cone below the lens focuses incoming visual stimuli into the next underlying organ, the rhabdom. The rhabdom is an elongated conduit that terminates at the basement membrane. The nerve axon beneath collects and feeds the information to the brain. Compound eyes sense motion far better than we do, but are poor at resolving detail. Still, paper wasps (*Polistes* spp.) recognize the faces of nestmates and behave submissively or dominantly accordingly. Many wasps perceive the ultraviolet end of the light spectrum, and can navigate by polarized light.

Heads Up
The head of a yellowjacket, *Vespula germanica*, is equipped with mandibles, palps, compound eyes, and antennae. Sensory hairs provide added input.

A trio of "simple eyes," called ocelli, is located in a triangular formation between the compound eyes. In some wasps these are reduced to scars, or absent. In nocturnal Darwin wasps like Ophioninae, the ocelli are strikingly large and conspicuous. The ocelli distinguish light intensity and may help insects orient to the horizon.

Protruding from the middle of the face are a pair of antennae. More than "feelers," antennae exemplify the chemotactile experience of insects. Delicate hairs and sensory organs constantly evaluate incoming chemical and textural stimuli. Is it the flavor of a potential meal? The fragrance of the opposite sex? Vibrations of an approaching predator? A yellowjacket may be fooled by a nail in your patio deck that is the same size as a fly. "Food!" says the eye. "Not food" says the antenna.

Mandibles—the jaws of the insect—work in opposition to chew food, manipulate nesting material, and bite adversaries. Above the jaws is the upper lip, the usually concealed labrum, and atop this is the clypeus, a trapezoidal plate below the antennae and between the eyes. The lower lip, called the labium, works in concert with the labrum to work food into the mouth. Jointed, finger-like mouthparts help orient food; these are the labial palps, attached to the labium, and the maxillary palps, joined to the paired maxillae. Maxillae, located between the mandibles and labium, aid the mandibles in cutting or grinding food.

The brain is not the only hub for neural transmissions. The subesophageal ganglion, for example, interprets and responds to stimuli related to the mouthparts, their muscles, and salivary glands. Other ganglia along the ventral nerve cord likewise act with some autonomy from the brain. Meanwhile, the aorta brings blood to the head, and the digestive tract transports food to the rear of the body.

Eyes Multiplied
The compound eye of a cuckoo wasp, *Stilbum cyanurum*, is composed of hundreds of individual facets. Although excellent for detecting motion, they do not resolve detail as well.

SPOTLIGHT ON

Antennae: Olfaction and Touch

Insects use their antennae much like we use our modern mobile devices, picking up information from the world around us. In wasps, antennae function differently in males and females, and are often modified accordingly.

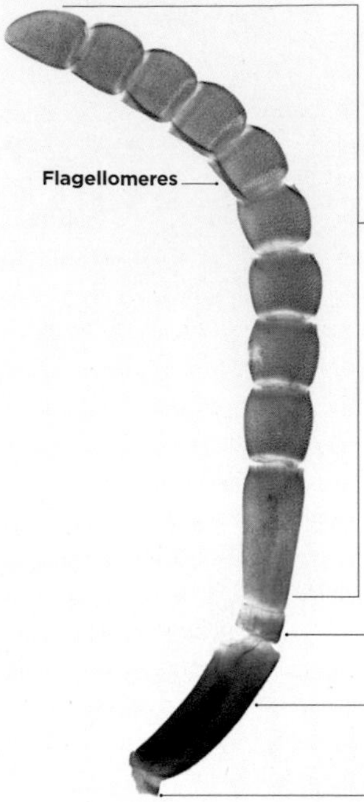

Sensory Wand
A wasp antenna is a series of segments outfitted with sensory hairs, pits, and other receptors for smell and touch. Males usually have more segments than females.

The wasp antenna is composed of three segments typically. The first is the scape, followed by the much smaller pedicel. The last segment, the flagellum, is longest of all, but divided into distinct articulations, each called a flagellomere. In the wasps most familiar to us, males have one more flagellomere than females. Muscles move only the scape and pedicel. Movement of the flagellomeres is achieved hydraulically, through changes in pressure of the hemolymph (blood).

In most wasps the antennae are filiform (thread-like), but they can be geniculate (elbowed), clavate or capitate (clubbed), serrate (saw-like), pectinate (comb-like), flabellate (fan-shaped), or plumose (like a bottle brush). Males of some mason wasps and paper wasps, have a distinct hook to the tip of each antenna. Antennae of male *Euceros* ichneumon wasps are sword-shaped. Some male Crabronid wasps, like *Rhopalum* and *Didineis* have clamp-like notches, presumably to latch onto their mate's antennae. Antennae of many male thread-waisted wasps (Sphecidae) have plate-like areas (placoids). Linear welts (tyli) occur on flagellomeres of some male Crabronidae. These bizarre modifications in male antennae reflect their primary function of locating the opposite sex by detecting her pheromone (scent), and grasping or stroking her during courtship.

Females use their antennae primarily to locate hosts for their offspring. This includes hammer-like organs in Orussidae, and other parasitoids of wood-boring insect larvae. The wasps "knock on wood" to generate and/or receive shockwaves that penetrate the wood and bounce off the host larvae or the walls of their tunnels.

Both sexes have hairs, pits, and other mechanoreceptors that inform the insect of tactile stimuli, air currents, flight speed when airborne, and substrate-borne vibrations. Chemoreceptors smell and taste the environment. So critical are the antennae that the insect constantly grooms them. Many wasps possess a "comb" for this purpose. The tip of the tibia of the front leg sports a long spur equipped with brush-like hairs, and the basal segment of the tarsus is concave opposite the brush. Each antenna is drawn through the comb for cleaning.

Hooks, Clubs, Branches
Modifications to antennae include hooked tips for grasping the antennae of a female (male mason wasp, right), and clubs and branches that create a greater surface area for sensory receptors (top, far right, and below). In most wasps the antennae are filiform (thread-like), but they can be geniculate (elbowed), clavate or capitate (clubbed), serrate (saw-like), pectinate (comb-like), flabellate (fan-shaped), or plumose (like a bottle brush).

Thorax: Locomotion Central

The second major body region, the thorax, is joined to the head by a narrow link called the cervix. This connection is elastic, but with at least one hardened ventral plate (sclerite) protecting the "throat." Through the cervix passes the front of the digestive tract, aorta, and ventral nerve cord.

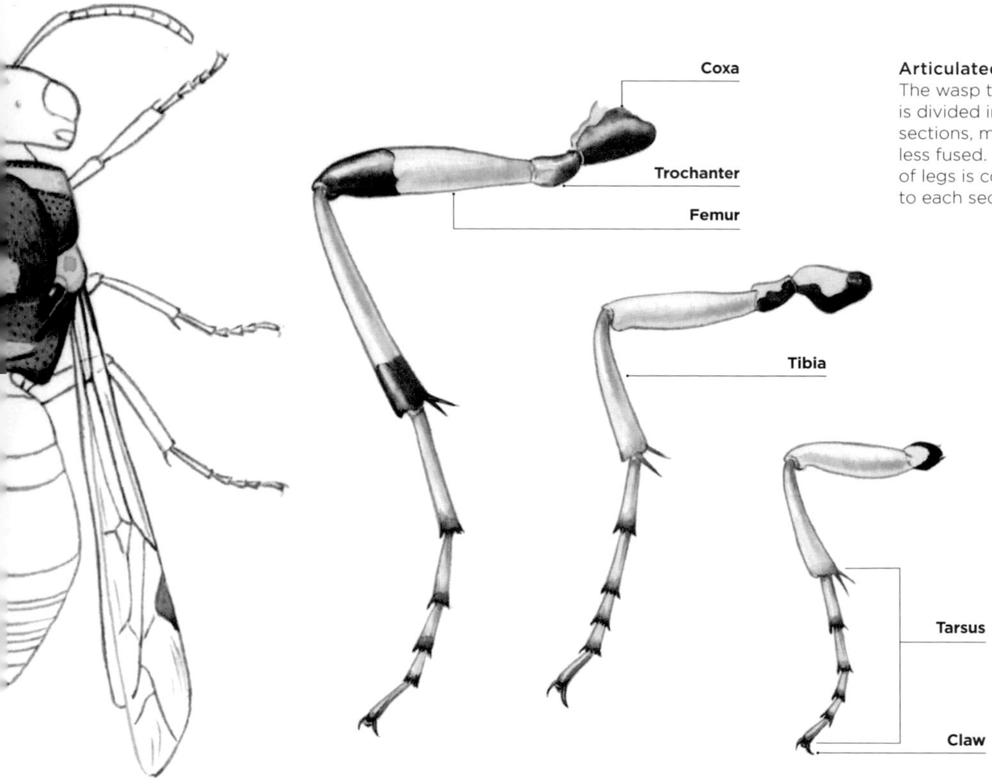

Articulated
The wasp thorax is divided into three sections, more or less fused. One pair of legs is connected to each section.

The thorax is itself a fusion of three segments. The front is the prothorax, the middle the mesothorax, and the rear the metathorax. One pair of legs is connected to each of these sections. There are further subdivisions in the form of exterior plates delineated by seams called sutures, ridges called carinae, and grooves or furrows called sulci. One spiracle (breathing hole) is located in the prothorax where the pronotum (top of first thoracic segment) meets the lateral (side) surface of the mesothorax. This hidden spiracle serves as the oxygen intake and carbon dioxide output for the insect's head, too. A second spiracle exists in the metathorax, often concealed by the base of the hind wing.

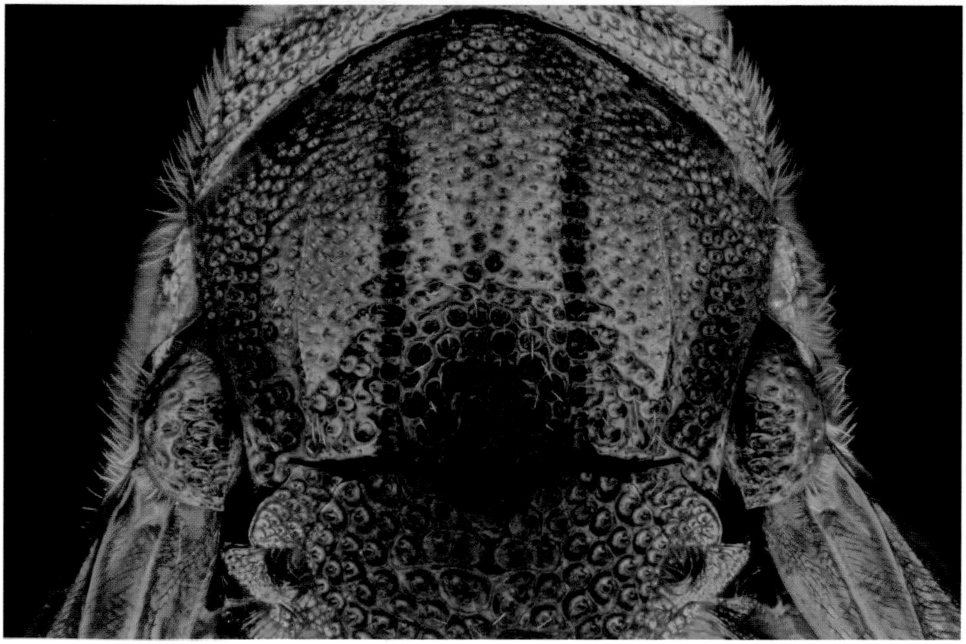

Armor Plating
The thorax of this cuckoo wasp, *Chrysis* sp., shows some of the distinct plates that compose it. Arcing over the front is the pronotum. Behind that is the mesoscutum. The "shoulder pad" at each wing base is the tegula. The scutellum is posterior to the mesoscutum.

True wasps, as well as bees and ants, have a unique feature that contradicts our visual perception of where the thorax ends and the abdomen begins. The rearmost segment of the thorax, called the propodeum, is actually the first dorsal segment of the abdomen, fused to the adjoining plates of the thorax. Because of this, entomologists usually call the *perceived* thorax of wasps, bees, and ants the "mesosoma." The articulation of the abdomen, the "wasp waist," therefore represents the second dorsal and first ventral segment of the abdomen. The propodeum also bears a pair of spiracles. In sawflies, horntails, and their kin (Symphyta), the true abdomen is joined broadly to the thorax—no propodeum is required.

One of the many advantages of having an exoskeleton is that it offers more surface area for muscle attachment. Nowhere is this more obvious than in the thorax of an insect where muscles work legs and wings. Insects may be the strongest of all animals in proportion to their size, but they are strong *because* of their small stature, not in spite of it. Most insects have such little body mass that they require little muscle support. That means greater power can be applied to lifting another object. A spider wasp easily hoists its limp victim up a wall, thanks to this ratio. The human equivalent might be dragging a piano up a flight of stairs with your teeth. Please do not attempt that.

No Neck
The head of a paper wasp is nested in the front of the thorax, protecting a vulnerable, flexible, muscular tube surrounding the nerve cord and esophagus.

Abdomen: Metabolic Headquarters

Whereas the head of a wasp is its sensory, memory, and behavior-initiation hub, and the thorax is the transportation center, the abdomen contains vital organs related to breathing, blood-pumping, digestion and excretion, reproduction, and self-defense/subduing hosts or prey. Because the first segment of the abdomen is fused to the rear of the thorax in true wasps, what we perceive as the abdomen is more properly called the metasoma. So, the first dorsal segment of the abdomen is the propodeum, and the second dorsal abdominal segment is the first segment of the metasoma. This can be confusing even to scientists if they fail to define their terms.

Terga
This photo of a *Ropalidia ornatifex* shows the concentric dorsal abdominal plates collectively called terga. The underside has complimentary ventral plates called sterna. This arrangement allows for articulation of the abdomen.

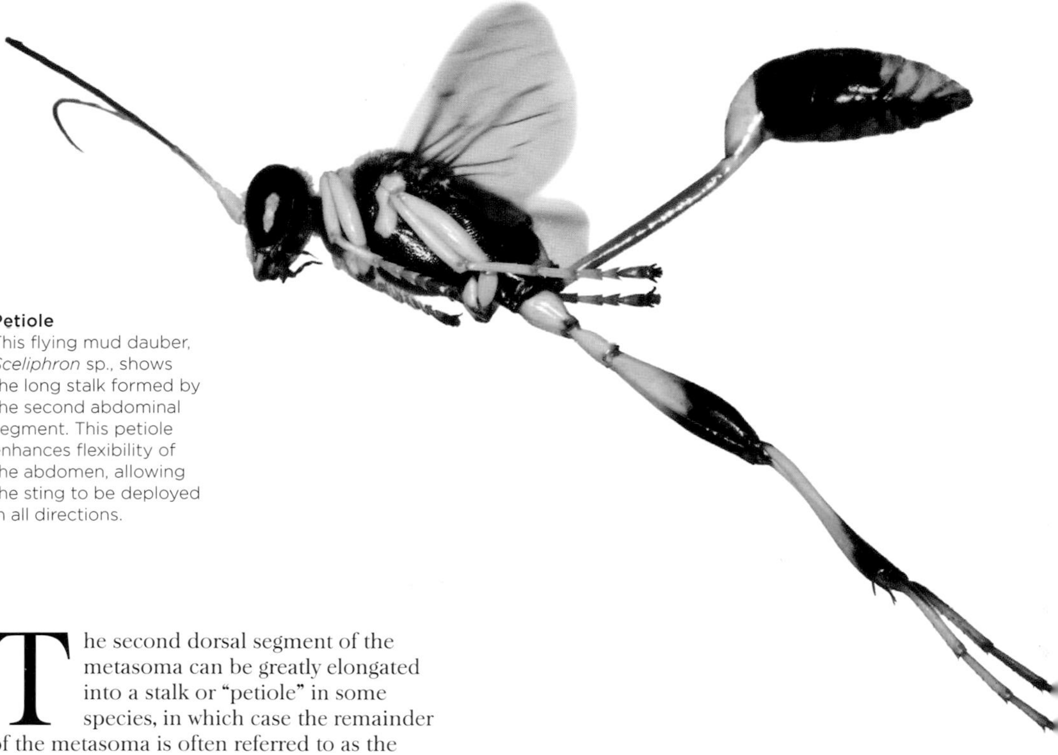

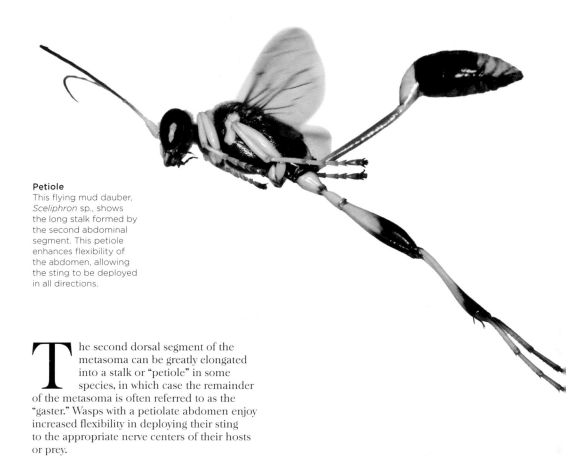

Petiole
This flying mud dauber, *Sceliphron* sp., shows the long stalk formed by the second abdominal segment. This petiole enhances flexibility of the abdomen, allowing the sting to be deployed in all directions.

The second dorsal segment of the metasoma can be greatly elongated into a stalk or "petiole" in some species, in which case the remainder of the metasoma is often referred to as the "gaster." Wasps with a petiolate abdomen enjoy increased flexibility in deploying their sting to the appropriate nerve centers of their hosts or prey.

Inside the metasoma (or abdomen in sawflies and their relatives), the midgut and hindgut of the digestive system occupy the most space. At the forefront of the midgut is the "crop," an accessory organ for food storage that is technically the last part of the foregut. In social wasps, especially, food is often regurgitated from the crop to feed larvae and adult nestmates. Strange, stringy organs called malpighian tubules originate near the front of the hindgut. They are the insect equivalent of kidneys, purging the hemolymph and working in concert with the hindgut to regulate the balance of water and salts while drawing out nitrogen-based waste. Insects in general are efficient at water conservation, so wasp feces are mostly solid, and eliminated through the anus at the tip of the abdomen.

The ventral nerve cord and associated ganglia run the length of the abdomen below most other organs. The aorta—more properly called the dorsal vessel—carries hemolymph pumped from the heart to be freely distributed forward and throughout the body. The blood carries little oxygen, but furnishes nutrients to cells, and is the liquid in which hormones and other chemical messages travel. Networks of trachea branch throughout the abdomen as they do in the thorax and head, taking in air through spiracles, two on each side of each of the first few dorsal abdominal segments (tergites).

Reproductive organs are also housed in the abdomen, and the ovaries of female wasps may dominate a good deal of the body cavity. Adjacent to the oviduct are venom ducts connecting venom glands to the sting, in those wasps that sting.

Wings and Flight

Wasps that have wings usually have two pairs. The front wing, the forewing in entomology vocabulary, is the longer of the two. The hind wing is much shorter, but may have an equal surface area depending on the species or even the sex. The front wing and hind wing are joined when the insect is flying. This union is accomplished by tiny hooks on the leading edge of the hind wing that lock onto the edge of the forewing. These hooks are called hamuli. Wasps are extraordinary aeronauts, and some are among the fastest flying insects. A few can hover, like male sand wasps (family Crabronidae, tribe Bembicini), while females of the aquatic fairyfly *Caraphractus cinctus* use their wings as paddles to row themselves underwater.

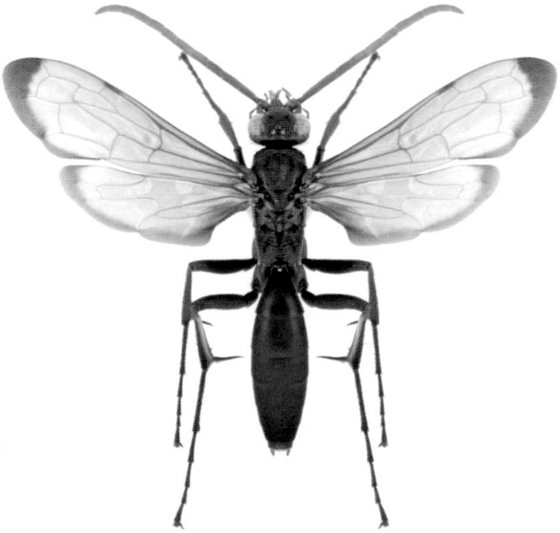

Surprisingly, the wings of wasps are nothing like those of birds or bats. The wings themselves have no muscle attachments. The entire thorax/mesosoma is a stunningly flexible box that is deformed by muscle groups contracting alternately. These asynchronous muscles cause the wings to go up or down, and the contraction of one group initiates the contraction of its opposing group once nerve impulses get the rhythm started. On the downstroke, the top of the thorax is depressed. Meanwhile, the side of the thorax bows slightly, accumulating elastic energy. At a critical point in the sweep of the wing that energy is released, helping drive the wing upward. This "click mechanism" enhances the power of the wing, and also contributes to faster wing movement. Wasp wings do not simply flap, though. They also rotate or twist to generate air currents that give the insect lift and thrust. Steering is accomplished by smaller muscle groups that flex the articulating segments that connect each wing to the thorax.

Membranes
The wings of this male tarantula hawk, *Hemipepsis brunnea*, show the veins of each wing, and the cells formed by those veins. Wing venation is often key to identifying wasps—at least to family.

Lift
This female digger wasp, *Sphex* sp., generates lift and propulsion with her two pairs of crystalline wings. She can also fly with a paralyzed katydid weighing at least as much as she does.

Hooked
This electron micrograph shows the row of hamuli that connect the hind wing to the forewing during flight.

Wing shape and configuration vary considerably. The forewing is folded longitudinally in social wasps when not deployed for flight. The arrangement of the wing veins, which lend structural support to the membrane, differ greatly from one family of wasps to the next, and also by genus in many instances. The network of veins also defines "cells," closed-off sections on each wing. Darwin wasps in the family Ichneumonidae, for example, are identified by the "horse head" cell in the middle of the forewing. Our scientific nomenclature for wing veins and cells has changed over time, and may vary depending on the preference of the individual entomologist describing them.

SPOTLIGHT ON

Ovipositor

Females of non-stinging wasps possess an egg-laying organ called an ovipositor. This takes many forms depending on the host plant or animal into which, or onto which, she is laying her ovum or ova. An ovipositor can be concealed within the abdomen, or protrude from its tip in spectacular fashion. The ovipositor of some wasps can be up to 12 times the body length of the insect.

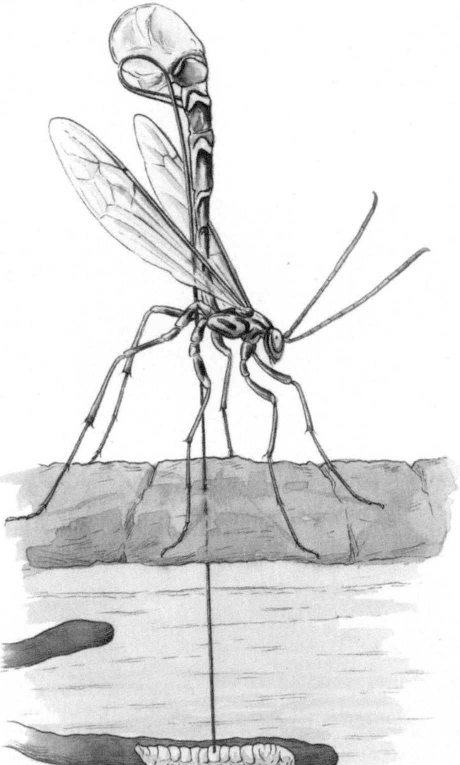

Insertion
This female sabre wasp, *Rhyssa persuasoria* of Austria (above), inserts her ovipositor into a tree where her host, a wood-boring insect larva, resides. The ovipositor is the thin filament, its sheath the thicker, bracing structure.

Drilling for a Host
A female *Megarhyssa* sp. (right) detects the presence of a horntail wasp larva (family Siricidae) within a tree by using her antennae. She then contorts herself to insert her ovipositor through the wood to reach her target.

The ovipositor is composed of abdominal segments modified into valves. The first gonaphysis is a divided, ventral valve complemented dorsally by the second gonaphysis, a single valve, to form the channel down which the egg travels from the oviduct. This arrangement allows more flexibility in movement than a single, hollow tube would.

Enveloping the ovipositor is the gonoplac, a sheath of two lateral, concave plates that form a protective sandwich over the delicate egg canal. This scabbard is often mistaken for the ovipositor itself, and when it splits it confuses the observer into believing something has gone horribly wrong. Dead wasps often exhibit this splintered arrangement. In life, the gonoplac is a remarkably strong structure that helps support the ovipositor during egg-laying.

It may appear that a wasp uses her ovipositor to "drill" into a host or substrate, but there is no circular motion as there is with our familiar power tools. Instead, the tips of one or more of the gonaphysis valves are toothed at the tip, and worked alternately and in linear fashion to achieve depth. In the aptly-named sawflies, much of the length of the ovipositor bears teeth that effectively cut into plant tissues.

The length of an ovipositor is a good indicator of its owner's host. Extraordinarily long appendages are usually associated with wasps that must penetrate a dense substrate to reach their host. This is true for parasitoids with hosts that are concealed, such as wood-borers in trees and logs, or the occupants of galls or figs. Short or retracted ovipositors are carried by wasps that usually contact exposed hosts, like caterpillars or aphids, directly. Many ichneumon wasps and braconid wasps fit in this category. These wasps may also deliver venom that temporarily paralyzes the host. After all, a violently thrashing caterpillar is not conducive to precision egg-laying.

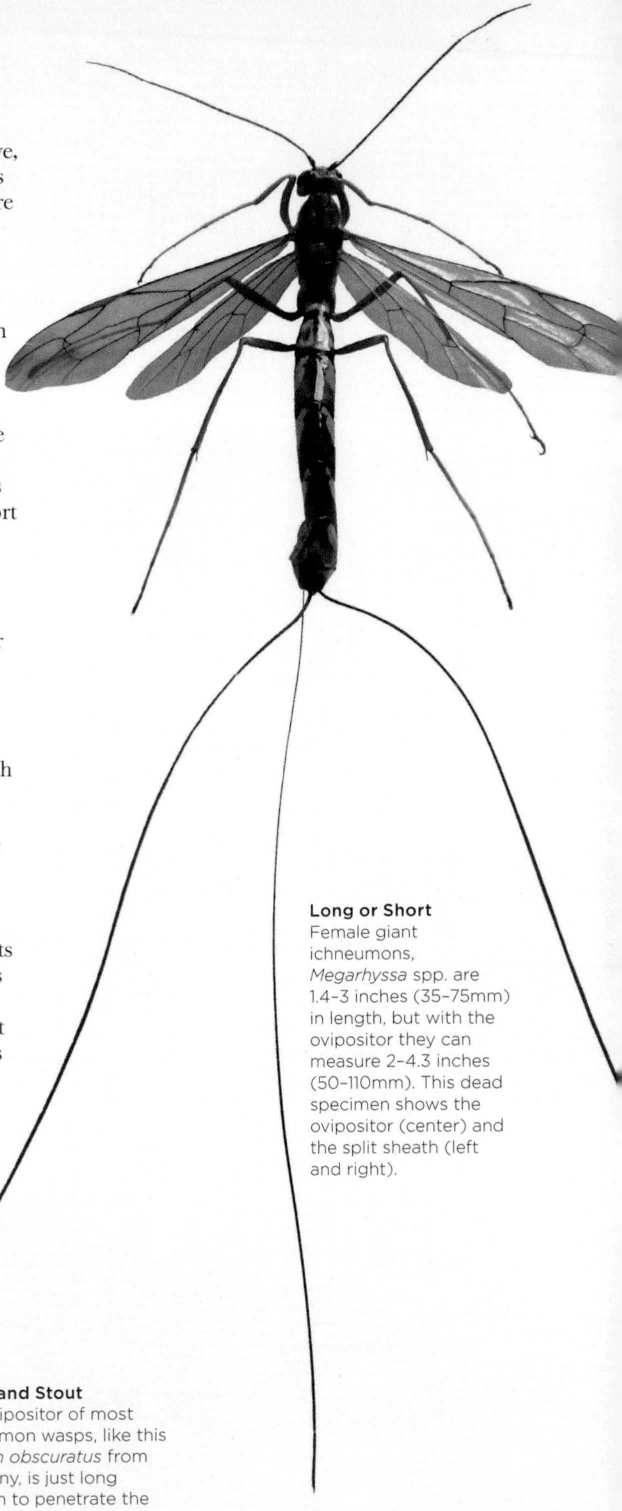

Long or Short
Female giant ichneumons, *Megarhyssa* spp. are 1.4–3 inches (35–75mm) in length, but with the ovipositor they can measure 2–4.3 inches (50–110mm). This dead specimen shows the ovipositor (center) and the split sheath (left and right).

Short and Stout
The ovipositor of most ichneumon wasps, like this *Ophion obscuratus* from Germany, is just long enough to penetrate the cuticle of the host insect or arachnid.

Sting and Venom

"Stinger" is a popular word for what is technically the "sting." It is unquestionably the greatest evolutionary achievement of wasps. What evolved initially as an egg-laying organ, the ovipositor, became weaponized. We are familiar with the sting as a tool of self-defense and colony defense in social wasps, but the primary purpose of the sting and its associated venom is to incapacitate struggling hosts or prey: inducing paralysis, temporary or permanent, is the goal. It is easier to transport, and/or lay an egg on, or in, a host if that animal is unable to flee, or offer resistance to the female wasp.

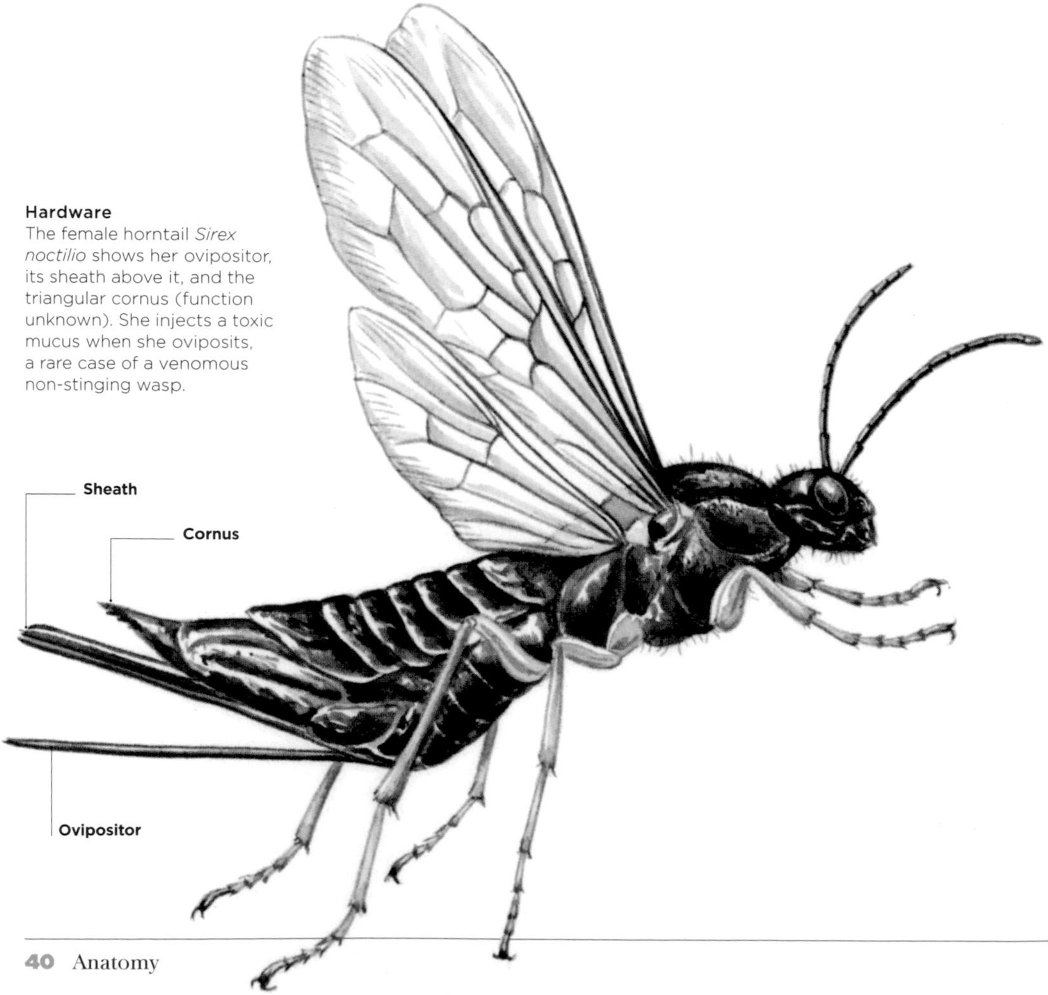

Hardware
The female horntail *Sirex noctilio* shows her ovipositor, its sheath above it, and the triangular cornus (function unknown). She injects a toxic mucus when she oviposits, a rare case of a venomous non-stinging wasp.

Venomous Needle
Normally retracted inside the abdomen, the sting of this female yellowjacket is here fully deployed. Above the sting is the exposed gonoplac, corresponding to the ovipositor sheath in non-stinging wasps.

The sting is usually retracted inside the abdomen of the female insect until deployment. Any visible filament, or spear- or needle-like extremity at the tip of the abdomen is either an ovipositor or, in the case of some male wasps, a pseudo-sting. The operative word is "usually." Nothing is ever simple in the wasp world.

Because the sting is derived from the ovipositor, it is valuable to compare the anatomy of the two. Neither is comparable to a hypodermic needle. What appears as a single structure is in reality composed of several parts that represent modifications of the terminal segments of the abdomen/metasoma. In a wasp ovipositor, two parallel, blade-like appendages form a channel down which eggs pass. Typically, the tips of each are serrated to facilitate entry into plant tissue or a host animal. They slide against each other alternately to advance the ovipositor into the tissues. The ovipositor is itself enveloped in a sheath composed of two halves of an additional appendage. The sheath filaments are stout, protecting the delicate ovipositor and serving as braces, like an oil-drilling derrick, if necessary, to penetrate a dense surface like wood.

A sting is built similarly to an ovipositor, but it lacks the sheath and is usually without teeth on the tips of the paired, piercing appendages. The result is a canal through which venom flows. Unlike the honey bee, which has barbs on the sting that anchor it in the wound, wasps can withdraw their sting for repeated use.

Venom is produced in one or more glands within the abdomen of the female. An increasing understanding of venom now defines it as basically any chemical or combination of chemical compounds that results in weakening or incapacitation of the host organism or victim. By that definition, the "non-venomous" horntail wasp, *Sirex noctilio*, envenomates conifer tree hosts with a cocktail that make the trees vulnerable to the fungus that is also introduced by the female wasp when she lays her eggs.

How is a Wasp Built to Sting?
The "wasp waist" allows multi-directional articulation of the abdomen to direct the sting wherever it is needed. The flexibility of the abdomen/metasoma/gaster itself allows the insect to contort its body to plunge its sting into struggling prey.

Top Ten Stingers

1. **Warrior wasp, Synoeca septentrionalis**
2. **Tarantula hawks, *Pepsis* spp.**
3. **Velvet ant, Dasymutilla klugii**
4. **Red-headed paper wasp, Polistes erythrocephalus**
5. **Red paper wasp, Polistes canadensis**
6. **Giant paper wasp, *Megapolistes* sp.**
7. **Yellow fire wasp, Agelaia myrmecophila**
8. **Glorious velvet ant, Dasymutilla gloriosa**
9. **Western yellowjacket, Vespula pensylvanica**
10. **Honey wasp, Brachygastra mellifica**

The most painful stings adapted from the "Schmidt Sting Pain Scale"

Sexual Dimorphism

Differences between male and female wasps are generally subtle, but the more practiced one becomes, the easier it is to identify each sex. Seasoned entomologists sometimes prank less-knowledgeable friends by freely handling male wasps that have no sting. Sometimes, the difference between males and females is so dramatic that one could be forgiven for assuming the two sexes are different *species*. Observable sex differences constitute sexual dimorphism.

Obviously, differences in genitalia exist between male and female wasps. Internally, males have testes, and females have ovaries. Male external genitalia are collectively the aedeagus, consisting of two penis valves framed by a pair of gonostyles, normally concealed in the abdomen. The corresponding female part is the vulva, the opening of the vagina. She is able to store sperm inside an internal pouch called the spermatheca. From it she releases sperm into her oviduct.

The Same But Different
Velvet ants, like *Nemka viduata* of Europe, demonstrate extreme sexual dimorphism. The male is winged, the female wingless. Their coloration differs as well.

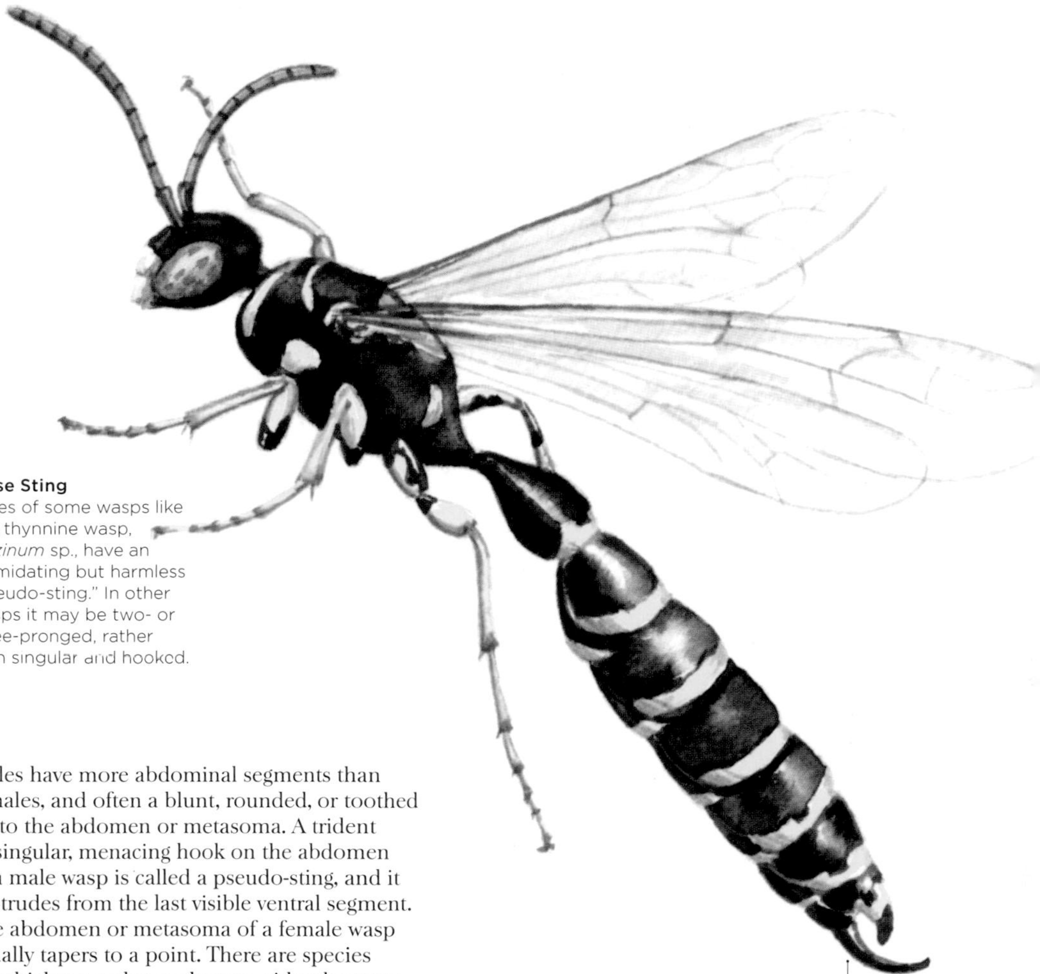

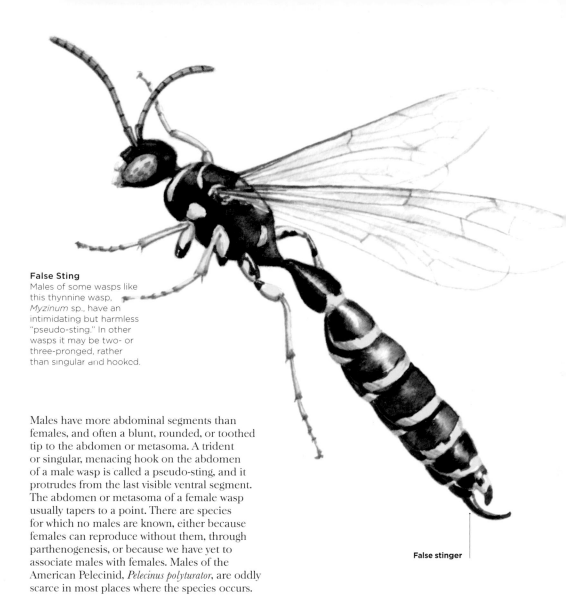

False Sting
Males of some wasps like this thynnine wasp, *Myzinum* sp., have an intimidating but harmless "pseudo-sting." In other wasps it may be two- or three-pronged, rather than singular and hooked.

Males have more abdominal segments than females, and often a blunt, rounded, or toothed tip to the abdomen or metasoma. A trident or singular, menacing hook on the abdomen of a male wasp is called a pseudo-sting, and it protrudes from the last visible ventral segment. The abdomen or metasoma of a female wasp usually tapers to a point. There are species for which no males are known, either because females can reproduce without them, through parthenogenesis, or because we have yet to associate males with females. Males of the American Pelecinid, *Pelecinus polyturator*, are oddly scarce in most places where the species occurs.

One of the most striking examples of sexual dimorphism is in the velvet ants, family Mutillidae. Males are usually macropterous (fully winged) and females are brachypterous (wingless). There is often a size difference, too, males being larger in some instances. This is even more pronounced in some members of the family Tiphiidae where wingless females are a fraction of the size of males, and a male can carry his mate on the wing. In-flight sex of this nature is called phoretic copulation. There are wasps, like fig wasps in the family Agaonidae, in which the male is wingless, but the female is winged. Still other species demonstrate winged, wingless, or vestigial-winged (micropterous) individuals of either sex, as in the genus *Trimorus*, family Platygastridae.

Antennae of male wasps typically have more segments than those of females. The male's antennae may also be modified in various ways to better detect the sexual scents (pheromones) that females deploy to attract mates. Male paper wasps in the subfamily Polistinae have hooked antennae, and often pale faces, making them easy to identify.

Bizarre Variations

Wasps can look weird, and we sometimes do not understand why. They can have horns, or spikes, or spines, or other modifications that may leave you scratching your head. Their legs can be expanded like the biceps of a bodybuilder, or the antennae can be clubbed or comb-like. The puzzling strangeness of such insects is part of their appeal to entomologists and naturalists.

A frequent reason for modification of the wasp body is to better accomplish the tasks associated with the lifestyle of that species. Wasps that excavate burrows for nesting are termed "fossorial," and the females of most species have a series of stout spines on the front legs called the "tarsal rake." Watching a sand wasp fling large quantities of soil under and behind her as she digs gives one an appreciation of the utility of her built-in tools. Curling her "toes" enhances the effect, putting all the spines in the same plane as she tunnels quickly into a dune.

Modifications to legs help males of some species to grasp the female during courtship and mating, and/or help the male guard his mate from competing suitors. Male shield-handed wasps in the genus *Crabro* have the tibia segment of the front leg expanded into an opaque shield that filters light into a species-specific pattern when placed over the eyes of the female.

Strange Things
This small wasp, in the family Eucharitidae, has comb-like antennae, the better to detect scents of the opposite sex, or ant nests. It also has long spines on the rear of the thorax, of unknown function.

Tarsal Rake
This female European beewolf, *Philanthus triangulum* (left), possesses long spines on her front "feet" that help her dig a burrow quickly and efficiently. Many wasps in the Crabronidae, Sphecidae, and Pompilidae families have this adaptation.

Tusks
Large male mason wasps in the African genus *Synagris* (below left) have bizarre horns on their jaws. They use these weapons against competing males while guarding nests from which a female is likely to emerge.

The faces of wasps can be adorned with a variety of structures. Males of some "tusked wasps" in the African genus *Synagris* have menacing horns sprouting from their jaws or face. They use these weapons primarily to battle for access to virgin females emerging from mud nests. They may also guard individual nests of females they have mated with, or patrol several such nests to defend against competing males. Males of the recently discovered "king wasp," *Megalara garuda*, family Crabronidae, from the Indonesian island of Sulawesi, also have exaggerated mandibles. We know nothing of their function. Female "warthog wasps" from Africa, *Genaemirum phacochoerus*, in the family Ichneumonidae, have odd projections on their "cheeks," perhaps helpful in bulldozing through the tunnels of their hosts: pupae of carpenterworm moths in the family Cossidae. The enormous compound eyes of male wasps in the subfamily Astatinae, family Crabronidae, meet at the top of the head—all the better to watch for females passing by.

Even the abdomen of a wasp can be bizarre. Take the "handle" on the front of the abdomen of tiny female wasps in the genus *Inostemma*, family Platygastridae. The projection serves to house the long, retractable ovipositor she uses to lay eggs in galls made by gall midges.

Ventral Spine
Males of the organ pipe mud dauber, *Trypoxylon politum* (below), have a menacing hook that they use to anchor themselves in nests that they guard from predators, parasitoids, and competing males.

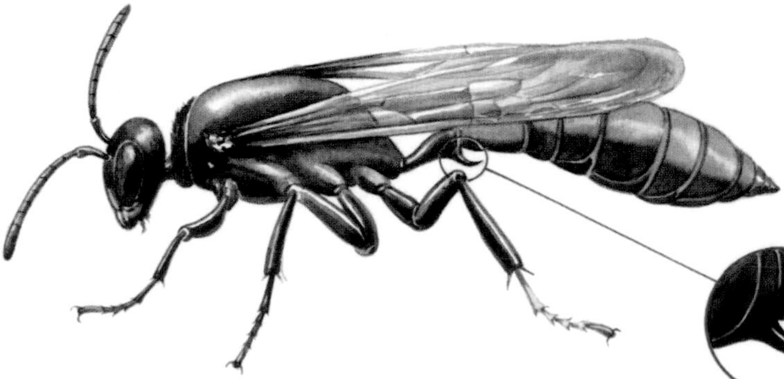

PROFILE

Dryinid Wasps

Family Dryinidae

The family Dryinidae does not have a standard common name, but calling them "scissor-handed wasps" would be appropriate. These are small, ant-like wasps with prominent eyes, the females of most species possessing mind-boggling adaptations of their front feet.

Family Dryinidae

SPECIES	~1,100
DISTRIBUTION	Worldwide except for Antarctica
SIZE	0.08–0.43 inches (2–11mm)
AMAZING FACT	Females of many species have pincers

Actual size

Small but Mighty
Adult dryinid wasps average around 0.2 inches (5mm), and rarely exceed 0.43 inches (11mm). What females lack in size they make up for in ferocity.

Stuck on You
The larva of a dryinid wasp has formed a hardened sack for protection as it sucks the blood of its host, the nymph of a planthopper, *Siphanta acuta*, in Tasmania.

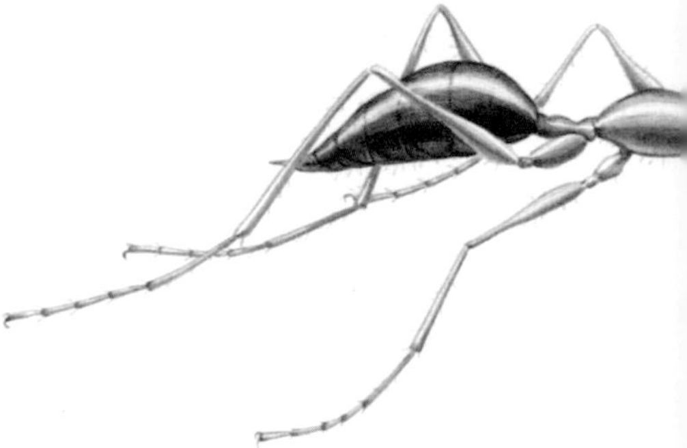

Females may be fully winged, but in a number of species are wingless (apterous), or have the wings reduced to non-functional pads (brachypterous). Males of most species have wings, a few being brachypterous, but all lack the intimidating hardware of the female's forelegs. Why the sexual dimorphism? Males do not wrestle host insects in hand-to-hand combat.

The scissor-like front tarsus of the female is a marvel of evolutionary engineering. One of the two claws is elongated into a blade-like appendage. The last (fifth) tarsal segment to which the claws are attached is expanded laterally to make the opposing "grip" against which the articulated claw compresses. Lacking a paralyzing sting, she uses these handy weapons to securely grasp a host and drive an egg between its abdominal segments. Nymphs and adults of leafhoppers in the family Cicadellidae, or planthoppers in the Delphacidae, Flatidae, and related families, serve as hosts for various dryinids. The adult female wasp will also kill hosts on occasion to feed herself, one of the few adult solitary parasitoid wasps known to engage in such predatory behavior.

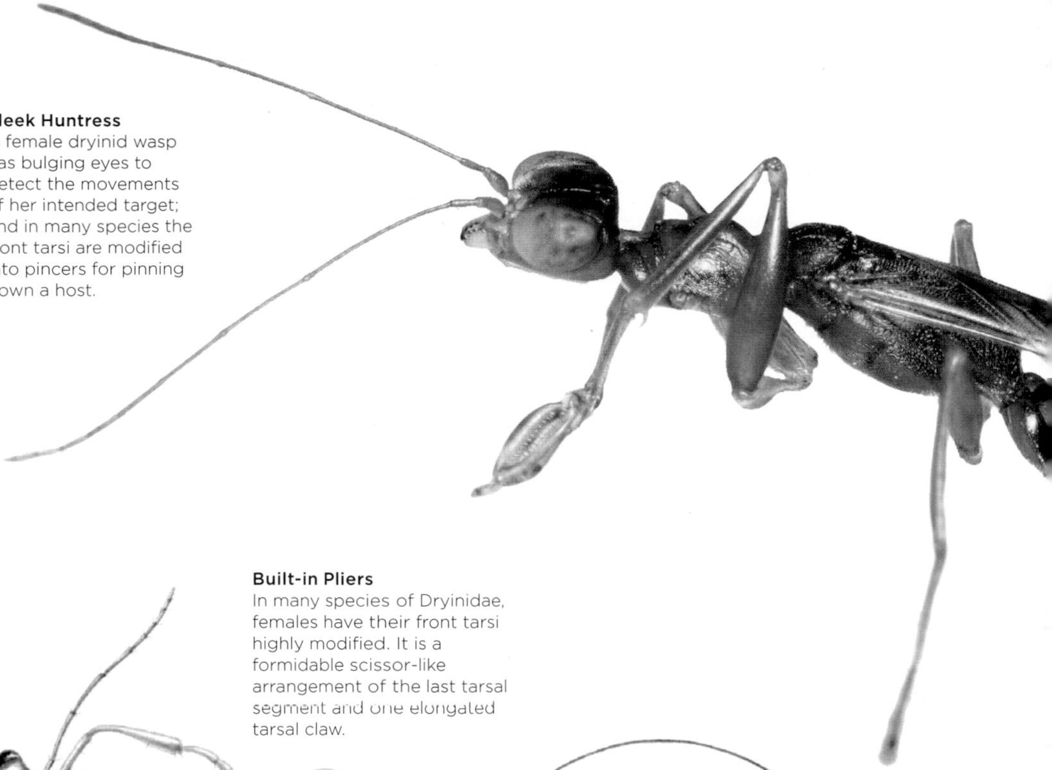

Sleek Huntress
A female dryinid wasp has bulging eyes to detect the movements of her intended target; and in many species the front tarsi are modified into pincers for pinning down a host.

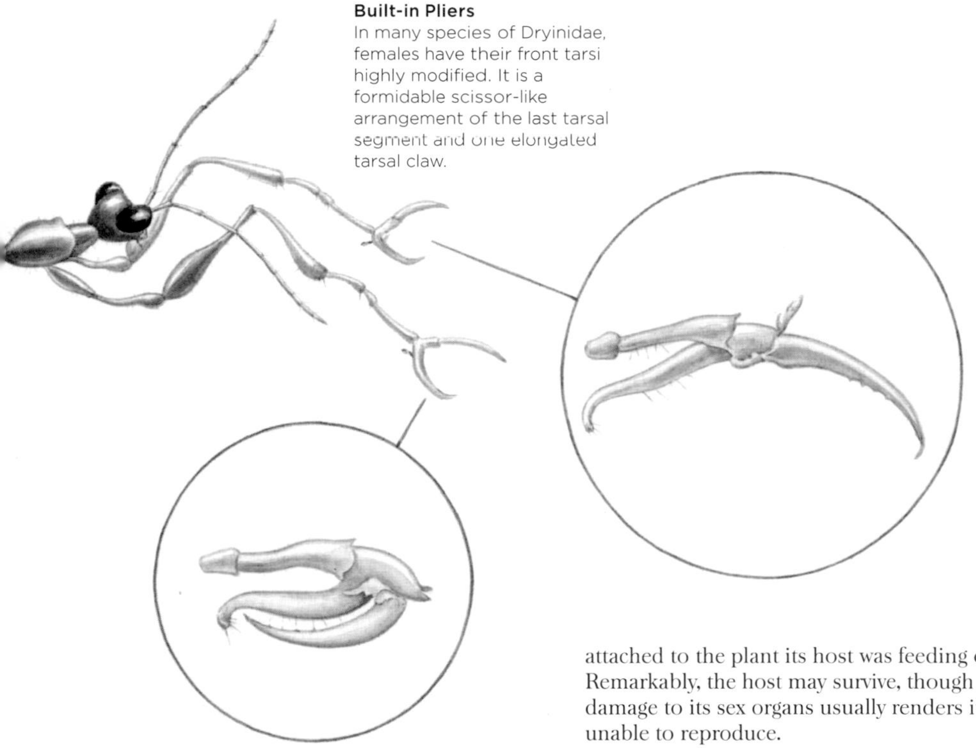

Built-in Pliers
In many species of Dryinidae, females have their front tarsi highly modified. It is a formidable scissor-like arrangement of the last tarsal segment and one elongated tarsal claw.

The dryinid larva that hatches from the egg develops initially as an internal parasite, but as it grows it eventually bursts the seams of its host, its hindquarters oozing between the host's body segments. The soft body of the larva is protected by a hardened sac called a thylacium, formed from its molted exoskeletons. When finished dining, the dryinid larva releases itself to enter the pupa stage in the soil, or attached to the plant its host was feeding on. Remarkably, the host may survive, though damage to its sex organs usually renders it unable to reproduce.

In rare instances, a single egg laid by the mother wasp will split into multiple, viable embryos (polyembryony, page 52). Under this scenario, the larvae remain internal parasites for the duration of their youthful lives.

Dryinidae are collectively found all over the world, with roughly 1,100 species in 58 genera and 11 subfamilies.

3

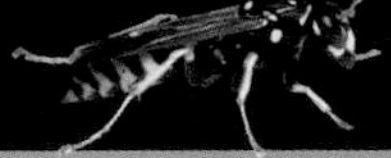

Metamorphosis
Purpose in Transformation

The Wasp Life Cycle

All wasps are holometabolous insects. That sounds frightening, but all it means is that they undergo complete metamorphosis. This is a complicated, advanced life cycle shared by ants, bees, beetles, flies, butterflies, moths, scorpionflies, caddisflies, lacewings and their kin, and fleas. Grasshoppers, cockroaches, mantids, true bugs, and other insects experience gradual or simple metamorphosis, with their juvenile stages growing incrementally and attaining a reproductive system—and maybe wings—with the last molt into adulthood.

The miraculous passage of wasps through egg, larva, pupa, and adult stages confers a number of advantages over simple metamorphosis. Each stage in the life cycle can be dedicated entirely to one or two functions. The larva stage can feed on one resource while the adult feeds on another, reducing or eliminating generational competition. One stage may excel over other stages in withstanding environmental extremes such as cold winters or hot summers. Complete metamorphosis usually results in faster growth than simple metamorphosis. Yet, there is more fluidity in the process than the four distinct stages would suggest. The final imago is almost literally imagined in the immature phases. Hormones play a pivotal role in all of this.

There are seemingly infinite variations on the theme of metamorphosis. Some wasp species that form plant galls alternate generations between a sexually-reproducing one and a generation without males. Female parasitoid wasps of some species are able to lay eggs that habitually split into dozens, hundreds, or even thousands, of additional embryos. A few wasps that are parasites of parasites undergo radical transformations within the larval stage in a phenomenon known as hypermetamorphosis, a strategy for seeking elusive hosts. You are unlikely to see a wasp egg, larva, or pupa because they are usually concealed underground, in plant tissue, inside another insect, in a paper or mud nest, or in a cocoon. It may come as a surprise to learn that a wasp may live the bulk of its life in one of those immature stages instead of as an adult.

Species that have only one generation per year are termed "univoltine." Other species may be bivoltine or multivoltine, with two or more generations each year. The duration of a life cycle for a given individual may vary even more. Wasps that feed as larvae inside plant seeds may be dormant for two or three years if conditions are not favorable for their emergence as adults.

An utterly unique feature of the order Hymenoptera is the manner in which sex is determined at birth. Fertilized eggs become females, and unfertilized eggs yield males. Consequently, females have two sets of chromosomes, males only one. This is haplodiploidy, a puzzling attribute that continues to confound entomologists.

Egg
An adult female social wasp, *Ropalidia* sp. in Thailand (above), has laid an egg in one of the cells of her paper nest. Within the egg, a yolk furnishes nutrition for the developing larval embryo.

Pupa
The pupa of a bethylid wasp (family Bethylidae) (right) is revealed inside a kernel of grain damaged by its host, the lesser grain borer beetle, *Rhyzopertha dominica*. The wasp helps keep this pest in check.

The Egg: Embryogenesis

The ovum is the incubation stage of a future wasp, its sex already determined by its mother. Female wasps are able to store sperm in an organ called the spermatheca, and release the male sex cells at will. Eggs pass down the oviduct and, if fertilized, are destined to become female wasps. Unfertilized eggs will be male wasps.

Comb Home
A colony founding female paper wasp, *Polistes metricus*, is guarding the eggs inside the cells of her young nest.

Survival of this most vulnerable stage in the life cycle usually hinges on the mother wasp placing her eggs in the optimal location or circumstance for the next stage: the larva that will hatch from the egg. Sawflies and gall wasps must oviposit in the correct host plant. Parasitoid wasps must divine the identity of a host, as well as its location, sometimes concealed deep inside a tree trunk. Solitary stinging wasps lay a single egg upon paralyzed or deceased hosts they conceal in a cavity, burrow, or mud nest, or in some instances lay an egg in an empty cell before hosts are harvested. Social wasps lay one egg in each cell in the paper or mud comb. Where necessary, the eggs are cemented in place with secretions from Dufour's gland, an organ inside the adult female's abdomen.

There are plenty of exceptions to straightforward scenarios. Some wasps "broadcast" scores of eggs, scattering them in places where the host organism is likely to come to them, instead of the female wasp searching for victims. In such cases the host usually ingests the leathery or hard-shelled eggs, which then hatch inside the host. Few offspring will reach the proper destination this way. The opposite approach is to lay fewer eggs, but protect them. Some sawfly species invest in parental care, the mother wasp fiercely guarding her clutch of eggs, and the young larvae that hatch, against predators and tiny parasitic wasps.

Another extraordinary strategy is polyembryony. Members of four families of Hymenoptera are able to inject a single egg into a host and then have the embryo split into many more: upward of 2,000 additional embryos in some cases. If this sounds somewhat familiar it is because this is the same phenomenon that produces twins or multiple human babies per single conception.

Subduing another organism, even another insect or a spider, is fraught with danger, so many wasps evolved venomous stings as a way to make the host more...compliant. This usually translates to complete paralysis—temporary or permanent—so that attaching or injecting an egg is an easier task.

Ovipositing
A female Krug's sawfly, *Sericocera krugii*, lays her eggs on the leaf of a sea grape, *Coccoloba uvifera*, in Puerto Rico.

Ova
The eggs of a Krug's sawfly are opalescent red, indicative of a foul taste or poisonous nature. The mother wasp must lay her eggs on the proper host plant or her larvae will starve.

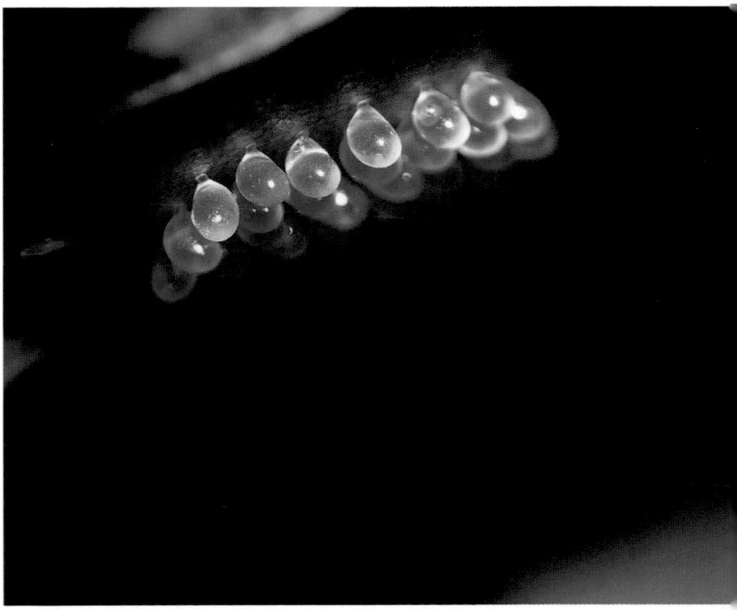

SPOTLIGHT ON

Rose Sawflies

Arge ochropus is one of several sawfly species that go by the name "rose sawfly," exposing one of the problems with common, English names. As one might expect, rose sawflies are considered pests due to their larval appetites for these ornamental garden shrubs. Still, they are an example of the life cycle of sawflies in general.

The female of *Arge ochropus* lays 15–18 eggs in the host plant by using her toothed, blade-like ovipositor to saw a groove in a stem or shoot and inserting her ova inside. She can lay around 35–50 eggs total during her lifespan, according to one laboratory study. The oviposition scar may become blackened and distorted. Early instar larvae feed together on the underside of leaves, scraping the lower epidermis of the leaf and/or flower buds. Later instars disperse to feed singly, chewing through leaves at the edge. Larval development through five instars takes about 24 days. Mature larvae measure 25mm, and drop to the ground to pupate in brown, double-walled cocoons they form in the soil. Pupation lasts about 14 days under laboratory conditions.

Third and fourth instar larvae are well-defended, and advertise the fact with a bright yellow head and broad dorsal stripe that contrasts with numerous black spots. Each abdominal segment is covered with 226 individual bristles that function as mechanoreceptors to detect the slightest touch. Should an ant or other small predator contact the spines, the larva suddenly raises its abdomen to expose seven ventral glands, one per segment, that emit repellent, volatile chemicals. These alone may send the ant running, but if the antagonist persists, it may imbibe worse toxins that leave it nearly paralyzed, staggering away in retreat.

Native to Europe and Asia, including the Middle East and western Siberia, *Arge ochropus* usually has one generation each year. First-generation adults are seen in late April and early May, with second-generation adults appearing in July and August. The adult female is 7–10mm in length, and both sexes feed on pollen, especially that of umbelliferous flowers.

Damage to roses can harm growth and flowering; skeletonizing of leaves is common. This species has recently become established in Canada (Ontario, Quebec) and the northeast U.S.A. (Vermont, Massachusetts, New York, and Michigan).

Colorful
The adult rose sawfly, *Arge ochropus*, is a lovely insect. Its antennae have only three segments each, the last one a long, undivided segment. This is a trait of all argid sawflies.

Chewing Down
The larva of the rose sawfly munches greedily on a rose leaflet. It adopts a defensive posture here, to warn that it can secrete a foul substance if further antagonized.

Stingless Vegetarian
Rose sawflies may mimic in color and shape wasps that can sting.

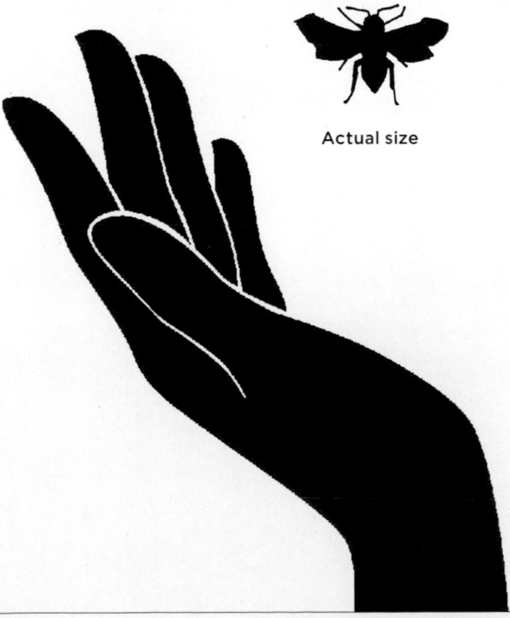

Actual size

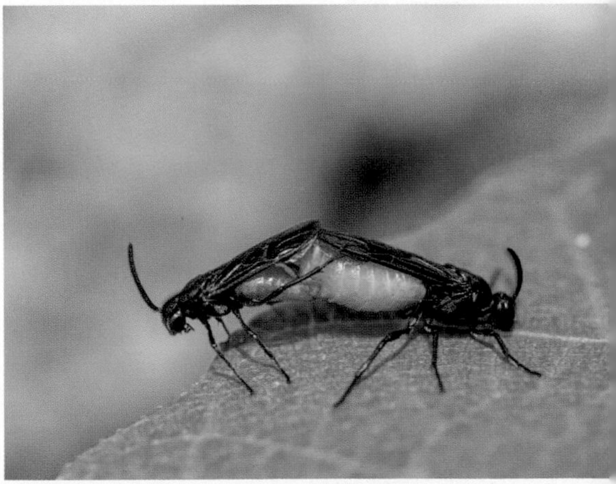

Mating
Male and female large rose sawflies *Arge pagana*, copulate back-to-back. Different wasps have different orientations during mating, depending in part on the configuration of their genitalia.

The Larva: An Eating Machine

Teenagers are the eating and growing phase of the human lifetime, and the larva stage is the equivalent for wasps. The larva usually represents the greatest percentage of time in the total life cycle, and often includes a prepupa stage that undergoes diapause to endure winter. Diapause is characterized by dramatically reduced metabolism, and temporarily arrested development. Dedicated as they are to feeding and growing, wasp larvae generally sacrifice the mobility and armor of the adult.

Larva and Food
Contents of a nest of the grass-carrier wasp *Isodontia mexicana*, dumped on a table, show the larva amid a cache of paralyzed katydids that it is consuming. Black specks are the larva's frass (faeces). Nests of *Isodontia* are in linear cavities partitioned with grass.

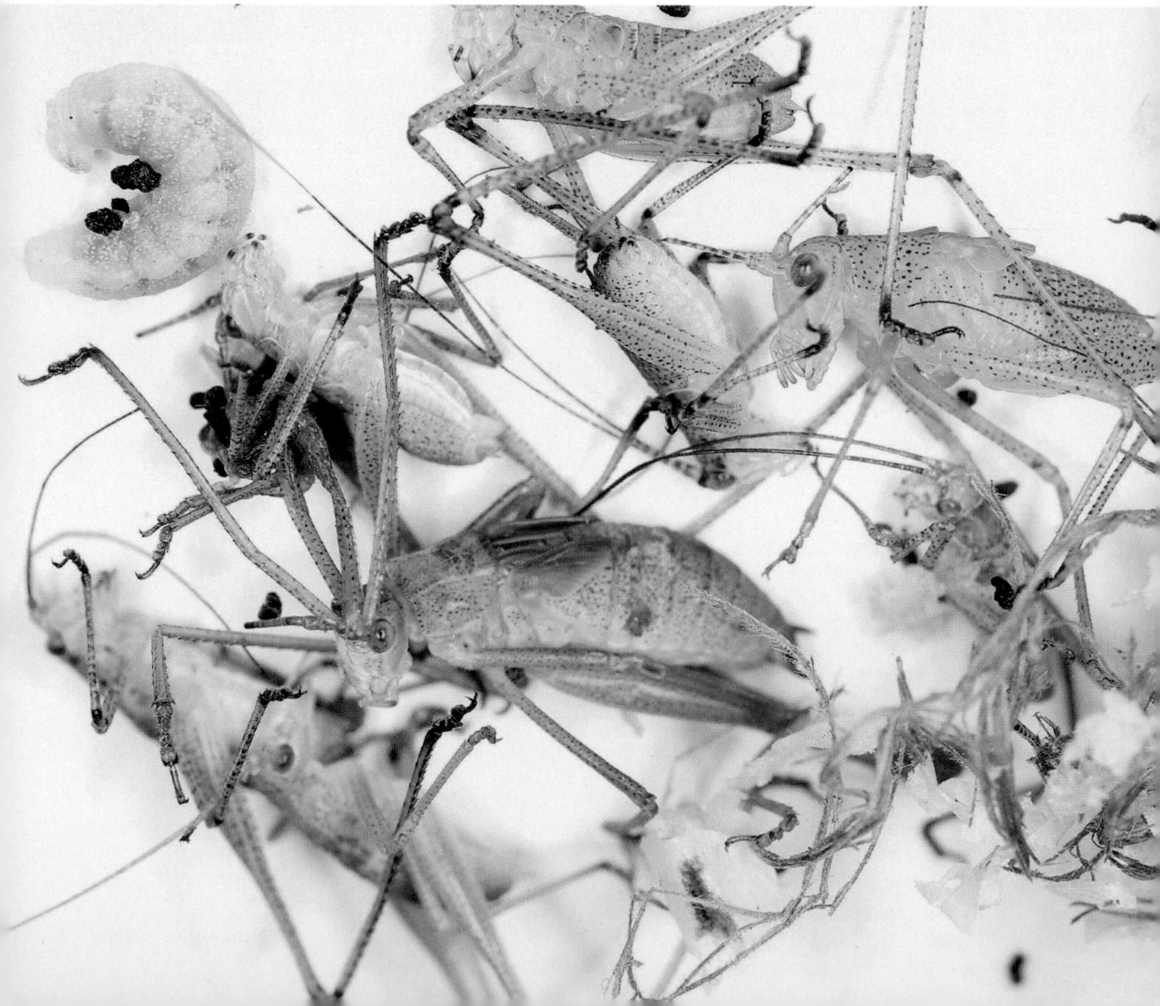

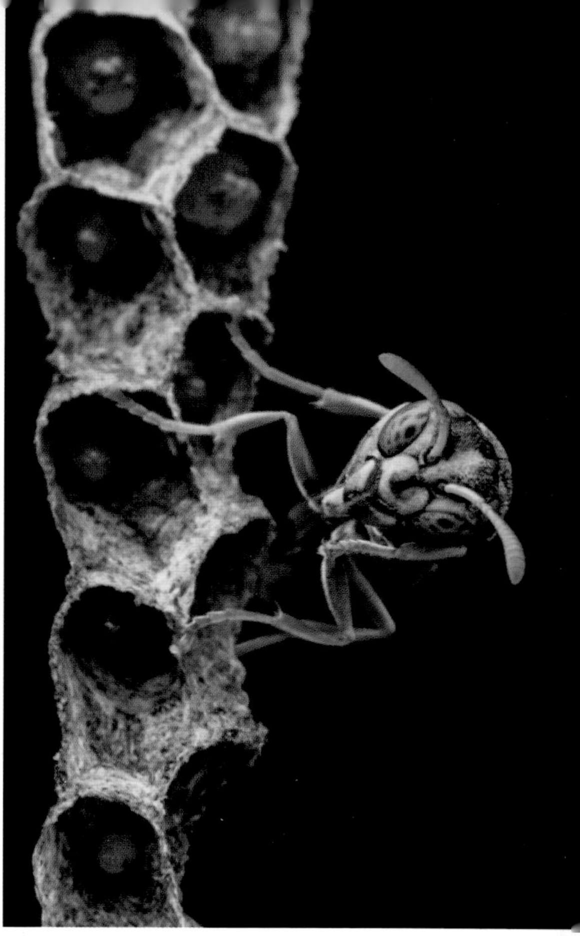

Vulnerable
The larval members of a colony of social wasps, *Ropalidia fasciata*, are protected by an adult female. Each larva is ensconced inside a paper cell where its only responsibility is to eat and grow.

Despite their soft, flexible bodies, there is a limit to the elasticity of the larval exoskeleton. The larva must molt periodically, shedding the old exoskeleton and expanding before the new one hardens. Intervals between molts are called instars. There are an average of 3–5 instars, but the number and duration of each varies according to sex, amount of food available, environmental conditions, and species identity. Molting (ecdysis) and growth are regulated by hormones, the most important being juvenile hormone or "JH." JH maintains the expression of juvenile characteristics but, inside, the blueprint for an adult wasp is already in place. "Imaginal discs" destined to become wings, legs, compound eyes, and antennae are triggered by other hormones.

Foods high in protein fuel growth, but sawflies and their kin, and gall wasps, consume cellulose as larvae. Securing enough nutrition from plants necessitates eating large quantities, and digesting it with the help of symbiotic microbes in the gut of the insect. Most wasp larvae feed on other invertebrates. The food source (host) is living, paralyzed, or deceased, depending on the wasp species.

Most parasitoid wasp larvae live on or within insect larvae, insect eggs, or spiders. A few are hyperparasites: parasites of other parasites that must find a host within a host, say a braconid wasp larva inside a caterpillar. Hyperparasites undergo hypermetamorphosis whereby the larvae change drastically. The family Perilampidae includes such examples. The first instar perilampid larva is a planidia, an active, host-seeking missile. Once a host is located, it attaches, and the succeeding instars are plump couch potatoes by comparison.

Stinging parasitoid wasps are provided for by their mother. She collects at least one host, usually an insect or spider, paralyzes it, and caches it in an underground burrow, pre-existing cavity in wood or other material, or inside a mud cell. She lays a single egg, seals the nest, and departs. The larva that hatches from the egg consumes the food, which won't rot because it is still living. Social wasps kill prey outright, chew it up, and bring it back to the nest where the prey is distributed to the larvae living in the cells of the comb.

PROFILE

Paper Wasps

Polistes spp.

Provided you are brave, paper wasps in the genus *Polistes* offer a unique opportunity to witness a wasp's life cycle. The uncovered combs of these social insects are a window into metamorphosis, though a curtain of silk conceals the pupa. You may want to use binoculars, the zoom feature on your camera or phone, or a telescope, depending on your fear threshold. There are more than 200 species of *Polistes* on Earth, and their personalities vary.

Genus *Polistes*

SPECIES	>200
DISTRIBUTION	Worldwide except for Antarctica
SIZE	~0.3–1.5 inches (8–38mm)
AMAZING FACT	Inspired modern paper manufacturing

Actual size

Social Architect
Female paper wasps are all capable of reproducing, but only the dominant female (gyne) in a given colony does so. She bullies her cohorts into raising her offspring and expanding the nest.

Besides individual transitions, paper wasps have a colony life cycle. *Polistes* is Greek for "city founder," and one or more female foundresses establish a single nest. These are gynes, females that reproduce but do not differ in size or form from other females. One female eventually asserts dominance, and suppresses the development of functional ovaries in her daughters, who will be *de facto* workers. Nests are constructed of wood or plant fibers chewed into pulp and formed into paper. A thick pedicel suspends the comb from a branch, palm frond, or other surface. A small number of hexagonal cells is built, and a single egg laid in each.

Hard Worker
A female European paper wasp, *Polistes dominula*, performs many duties for her colony: she forages for prey, feeds other nestmates, feeds her larval siblings, helps build additions to the nest, and guards against predators and parasitoids.

Reaching Capacity
As it grows, the paper wasp larva fills the entirety of its confining hexagonal cell. Soon it will spin a silken dome over the opening in preparation for entering the pupa stage.

Transformation
Appearing inert on the outside, paper wasp pupae undergo radical changes internally. The closer they get to the time of emergence to adulthood, the darker the pupae become.

Newly Minted
The newest product from the paper wasp factory is an adult that chews her way through the silken cap that protected her in the pupa stage.

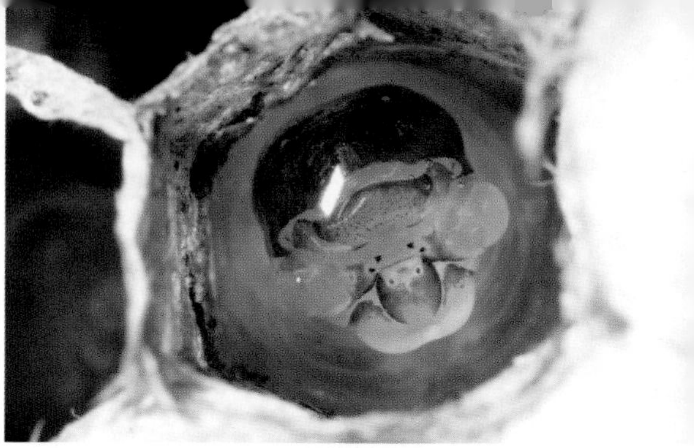

In temperate *Polistes* species, the foundress produces new gynes in late summer or autumn, along with males. In tropical species, this happens at other times, as dictated by wet and dry season variability in food abundance, and other factors. In *Polistes biglumis*, a high-elevation European species with a brief nesting window, females that lose their own nest may usurp the foundress of another nest. In fact, most nests of this species represent offspring from multiple females. *Polistes metricus* of the eastern U.S.A frequently re-uses old nests, maintains multiple active nests, and may share nests with other paper wasp species.

A study of *Polistes cinerascens* in Brazil found the duration of one colony cycle, from founding to producing reproductives, averaged 200 days with 94 adults produced. There were up to four generations per year with an egg stage that lasted 13 days. All larval paper wasps go through five instars before pupating; the larva stage lasts about 24 days in *P. cinerascens*. An adult emerges from the pupa in 22 days. The pupa stage of the temperate *Polistes dominula* lasts 10–14 days. Adult *Polistes* live about 38 days.

The Pupa: Epic Reorganization

The pupa appears inert externally, so is often referred to as the "resting stage" between larva and adult. Nothing could be farther from the truth. The pupa might better be termed the reorganizational stage, since that is what is transpiring internally. Like the larval stage, the pupa is usually hidden from view inside a nest or gall or plant stem, or within a cocoon, or in a cocoon *and* a nest. Resembling a ghostly adult, the pupa is pale initially, darkening as the adult insect takes shape. Nearly all wasp pupae are exarate, meaning the appendages are free from the main body. Many pupae in the Chalcidoidea are obtect, with appendages adhered to the body wall.

Bird Poop?
The pupa of the ichneumon wasp *Charops annulipes* is concealed within a silken black and white cocoon that resembles a bird dropping. The disguise protects the vulnerable pupa from parasitoid enemies.

Transitioning from lumpy larva to sleek, alert, active adult is no mean feat. Much can go wrong if the correct genes are not turned on, or off. Timing of hormone releases, and proper cellular responses to them, are critical. Juvenile hormone is no longer manufactured, signaling the cessation of juvenile programming and the end of larval anatomy, and the onset of construction of the adult central nervous system, and triggering of the imaginal discs into differentiating appropriate adult body parts.

Let us return to the larva for a moment, as it sets the stage for the pupa. Larvae of many wasps pass unfavorable environmental conditions as prepupae, and often protect themselves by spinning cocoons. Labial glands in the mouth of the larva excrete silk that the insect weaves around itself, helping to insulate it from excessive cold and offering a barrier to parasites. Once ensconced inside the cozy capsule, the larva may go into diapause, a state of torpor in which metabolism decreases drastically, and all growth and development ceases. By this point the larva has accumulated enough fat reserves to resume development on the far side of diapause. In the case of most social wasps, each larva simply spins a silken dome over the opening of its paper cell, and overwintering is accomplished by

Ghostly
The pale creature encased in a mud capsule is the pupa of a mason wasp, genus *Ancistrocerus*. This is an example of an exarate pupa: the appendages are free from the body of the insect.

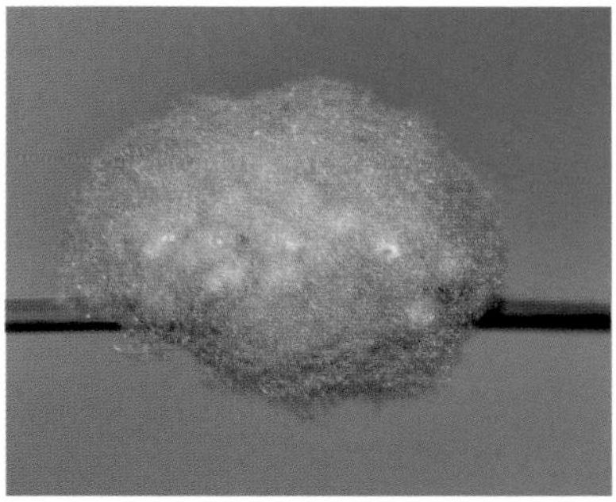

Fluffy
The cottony "cloud" on a pine needle represents communal cocooning by braconid wasp larvae of the subfamily Microgastrinae. The remains of their caterpillar host may lie underneath the woolly ball.

adult females. Some gregarious internal parasitoids (endoparasitoids) erupt from their host in unison, spinning small cocoons that may be mistaken for eggs. This is a common phenomenon observed on caterpillars.

Revving up the metabolism machine again, some larval tissue is destroyed in the course of creating the adult. Other cells are re-purposed, at the behest of hormones targeting specific gene locations. Once the process is complete, an adult wasp emerges from the pupa. It may need to chew its way out of a cocoon and/or a mud plug, or through a layer of plant matter, or some other barrier to its freedom. Not every wasp succeeds in breaking out.

Not Eggs
The "balloons" on this Carolina Sphinx moth caterpillar (*Manduca sexta*), are cocoons of a braconid wasp in the genus *Cotesia*. Dozens of wasp larvae have been feeding internally, and have finally erupted synchronously to pupate.

The Adult: Mating and Host-finding

The life of an adult wasp is all about finding a mate (males), procreating (both sexes), and providing for the next generation (females). It is also left to females to disperse to new or better habitats. There is much work to be done, but first to fuel up.

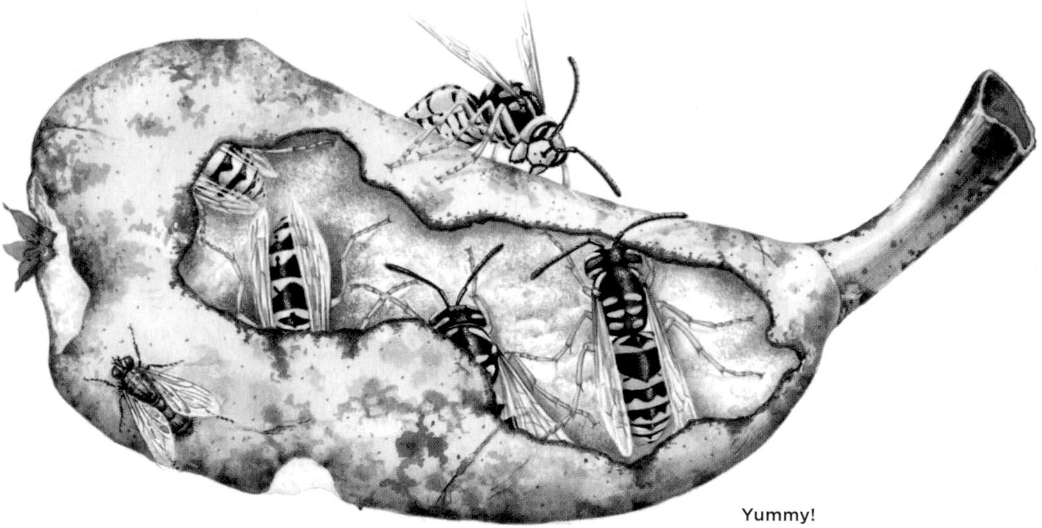

Yummy!
Adult yellowjackets, and a blow fly, feed on a rotting pear. Ripe fruit provides carbohydrates that the adult insects need for their bustling lifestyle. Fermenting? All the better.

Whereas the larva stage needed protein to grow, adult wasps require carbohydrates to stoke their metabolic furnace. Flower nectar is the wasp equivalent of our energy drinks, so you will often find wasps on blooming plants. When nectar is scarce, wasps avail themselves of alternatives. This includes sweets from extrafloral nectaries, specialized structures that produce nectar apart from flowers. Sunflowers (*Helianthus* spp.) possess them, for example. Plant galls may also exude nectar. Sap oozing from wounds on trees, all the better if fermenting, also suffices. Overripe fruit is another resource some wasps exploit. A favorite indulgence is "honeydew," the liquid wastes excreted by aphids, scale insects, planthoppers, treehoppers, and related true bugs that live in groups. Aphid colonies may attract scores of wasps of diverse families, as well as bees, flies, ants, and other insects.

In temperate climates, adult females of some species overwinter, and these we see early in spring. They include ichneumon wasps, paper wasps, yellowjacket queens, hornet queens, solitary cutworm hunters, and cricket-seekers. Having mated the previous autumn, they are now free to hunt and nest without being bothered by overzealous males.

Among solitary wasps, males are usually the first to emerge from nests, and they immediately seek virgin females. Strategies vary, but a primary method is to simply force themselves onto the opposite sex. A tumbling ball of male wasps wrestling over a female concealed somewhere inside their frenzied mass may injure one or more parties. Less chaotic approaches to mating include staking out a landmark where a male can see females approaching, and from where he can chase away competing males. Male wasps of some species establish territories, sometimes scent-marking like a cat. In a few species, males assist females *after* mating, guarding the nest while the female is gathering building material or hunting for prey.

It is a naturally human desire to see other animals as inferior to us, but wasps give us ample reason to call our superiority into doubt. Female wasps are capable of astounding feats of fearlessness, strength, navigation, and parental care. Their instincts maintain enough plasticity to allow them to overcome novel problems. They lack a mammalian level of intellect, yet they not only survive, they thrive, in a diversity of ecosystems.

Calling All Males
A wingless female flower wasp in Australia, family Thynnidae, emits a pheromone and waits for a much larger, winged male to sweep her off her perch. This is her only way to reproduce, get to a nectar source, and find her way to a more distant, favorable habitat.

Honeydew Addict
Many wasps, like this diminutive cuckoo wasp, *Pseudomalus auratus*, crave the sugary liquid waste (honeydew) produced in copious amounts by aphids, scale insects, treehoppers, other sap-sucking bugs, and even some kinds of galls.

PROFILE

Emerald Cockroach Wasp

Ampulex compressa

One of the shining heroes of the wasp world is *Ampulex compressa*, the emerald cockroach wasp. True to its name, it dispatches cockroaches of the genus *Periplaneta* in an especially macabre manner. The female wasp stings the roach in the thorax to stun it, then uses her jaws and upper lip (clypeus) as a clamp to seize the roach by its hood-like pronotum. She then stings it in a nerve center under its head. The venom acts to suppress her victim's ability to move voluntarily. She amputates the roach's antennae, drinking what fluid leaks out. Using the shortened antennae like a horseman uses reins, the wasp then guides the roach to an appropriate natural cavity that will serve as its tomb. Once inside, the wasp lays her egg at the base of the roach's middle leg. She then carefully plugs the entrance to the cavity with debris and leaves.

Family Ampulicidae

SPECIES	~170
DISTRIBUTION	Worldwide, mostly tropical
SIZE	0.2–1.3 inches (5–33mm)
AMAZING FACT	All species use cockroaches as hosts

Actual size

Roach-master
Ampulex compressa is the largest species in its family, but still smaller than its cockroach host, measuring 0.7–0.98 inches (18–25mm) compared to the roach's average 1.6 inches (40mm) length.

A Lip With Grip
The clypeus (face plate below the eyes) and jaws of *Ampulex* work like a clamp to grip the edge of a cockroach's carapace, so the wasp can sting it underneath.

Life Cycle
A female wasp has stung her cockroach victim into a cooperative state and drags it where she wants by its antennae. Another sting at the nest paralyzes it permanently. Her egg has hatched and the larva feeds externally, then internally. A new wasp emerges from the body cavity of its now deceased host.

Inside the nest, a larva hatches from the egg in three days, puncturing the cuticle of its still-living host, and begins to suck its blood. The larva does a remarkable thing in spitting out copious amounts of an antimicrobial cocktail that "disinfects" the roach prior to feeding. The larva spends two instars as an external parasitoid, using its piercing mandibles to feed.

Molting to the third instar, about seven days since it hatched from the egg, the larva now has more blunt jaws that it uses to tunnel inside the roach. This finally causes its death, as it consumes all muscle and internal organs, save for the digestive tract and Malpighian tubules. Once it has finished dining, the mature wasp larva spins a cocoon around itself, still within the empty husk of the roach. The entire egg-to-adult cycle takes around six weeks, and the adult female wasps may live for several months.

Averaging 0.86 inches (22mm) in length, the female emerald cockroach wasp is one of the larger members of the family Ampulicidae, formerly treated as a subfamily of the thread-waisted wasps (Sphecidae). There are roughly two hundred species in six genera. This species is found in Africa, Asia, Australia, and many tropical Pacific islands. Much smaller, duller species occur in North America, but Ampulicidae in general are overwhelmingly tropical, and all known species have cockroach hosts

Alternation of Generations

Interestingly, some gall wasps in the family Cynipidae practice heterogony. That is, they have a life cycle of alternating generations between a traditional sexual generation involving males and females, and an asexual generation of females that reproduce without males via parthenogenesis. This bizarre pattern is limited to gall wasps of the subfamily Cynipinae living on oak or sycamore trees, as far as we know. The genera *Andricus*, *Cynips*, and *Neuroterus* demonstrate the typical form of heterogony.

Yearly Cycle
Hedgehog gall wasps, *Acraspis erinacei*, produce spiny galls on oak leaves. Wingless, parthenogenic females emerge in late autumn, laying their eggs in buds. The following spring, male and female wasps emerge from the cryptic bud galls.

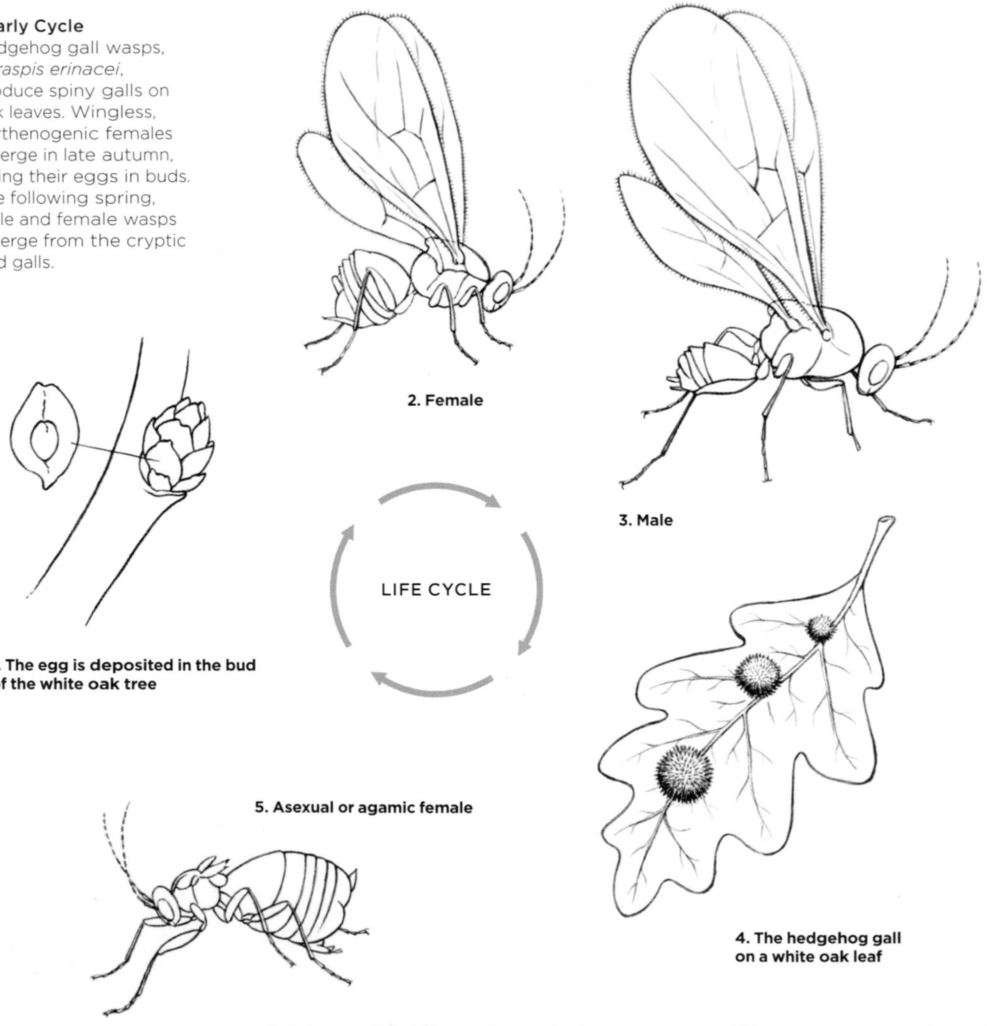

How is this possible? We know that haplodiploidy applies to all wasps: males come from unfertilized eggs, females from fertilized eggs. How can you get both males *and* fertile females without mating? You cannot have it both ways! Apparently, you can if you have two types of asexual females. Androphores are females that produce haploid eggs through meiosis. These eggs will result in males. Gynephores are females that produce diploid eggs that become sexually reproducing females. Asexual in this context means a female that reproduces without mating with a male (parthenogenesis).

One female, however, can only produce males or females, depending on whether she is an androphore or gynephore. Confused? It turns out that in polar opposite to all other wasps, these particular cynipids, in the *sexual* generation (females and males), cannot produce males from unfertilized eggs. Mating with males will produce only androphore females or gynephore females, depending on the individual (sexually reproducing) female. In conclusion, there are two types of females in each of the generations. Males are produced only by certain females of the asexual generation.

The galls most obvious to us are growths on the leaves or twigs or limbs of oaks. The insects inside represent the generation of asexual females only. These females are frequently wingless, and typically emerge in late autumn. They lay their eggs in the developing buds of the host. A tiny gall forms on the scales of the bud. In late spring or early summer of the following year, the sexual generation of males and females emerges from these bud galls. They are winged and disperse to start the cycle again. The asexual generation alone may require one year, or a year for each generation. This can vary by species and/or depend on environmental conditions.

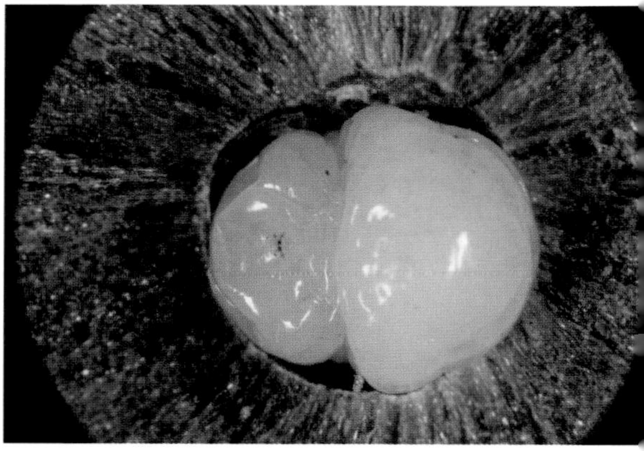

Cramped Quarters
This cross-section of an oak marble gall (right) shows the mature larva of the wasp *Andricus kollari*, in the central compartment, folded in half and about to pupate.

There are other deviations from the above, namely in the sycamore gall wasp, *Pediaspis aceris*, and some oak gall wasps, including *Biorhiza pallida* and some *Andricus* species. Some individual, asexual (parthenogenic) generation females of *Biorhiza pallida* can produce both females *and* males, for example. Lest your mind be bent no farther, take comfort in the fact that the genetic origins of heterogony are completely unknown, even to scientists.

Oak Apple
Unlike the dense, woody marble galls on stems, oak apple galls are on leaves, and spongy on the inside (below). A central chamber, suspended by a web of filaments, is where the wasp larva feeds.

4

Beauty
Purpose Expressed in Structural Design

Iridescence and Aposematism

Few people consider wasps to be beautiful creatures until confronted by their diversity. The endless variety of shapes, sizes, colors, and patterns serves the extraordinary array of lifestyles these insects lead. What passes for mere beauty to us is vital to the survival of the organism that wears the exoskeleton.

Hard to Miss
The bold colours of an eastern velvet ant, *Dasymutilla occidentalis* of North America, warn of her sting. Males are winged and do not sting, but they are seen less often than the conspicuous females.

The armor itself, regardless of color, serves several functions. Perhaps the most important is waterproofing. This has less to do with a terrestrial insect drowning than it does with that insect losing water through evaporation and transpiration. Spiracles—the breathing holes of insects—can be closed if necessary, but usually remain open. The exoskeleton also affords increased durability in the face of environmental wear and tear. Chitin, the major building material of insect armor, resists abrasion, splitting, and other forms of damage while discouraging predator attack. The cuticle of a cuckoo wasp or velvet ant may be exceptionally dense, even compared to other wasps, but it is still covered in sensilla, usually in the form of hairs. These setae detect changes in air movement, sense vibrations, or perceive chemical cues.

Many of the bright colors of wasps serve notice of their ability to defend themselves with a venomous sting, should circumstances call for such desperate measures. These visual advertisements are called aposematic colors or "warning colors." They may include iridescent colors like blue, especially if paired with a contrasting color. Metallic colors can be aposematic, oddly camouflaging (metallic green in some instances), or aid in thermoregulation. Wasps covered with patches of short, dense, silver or gold hairs reflect a great deal of incoming solar radiation, preventing the insect from overheating in extremely hot, dry habitats.

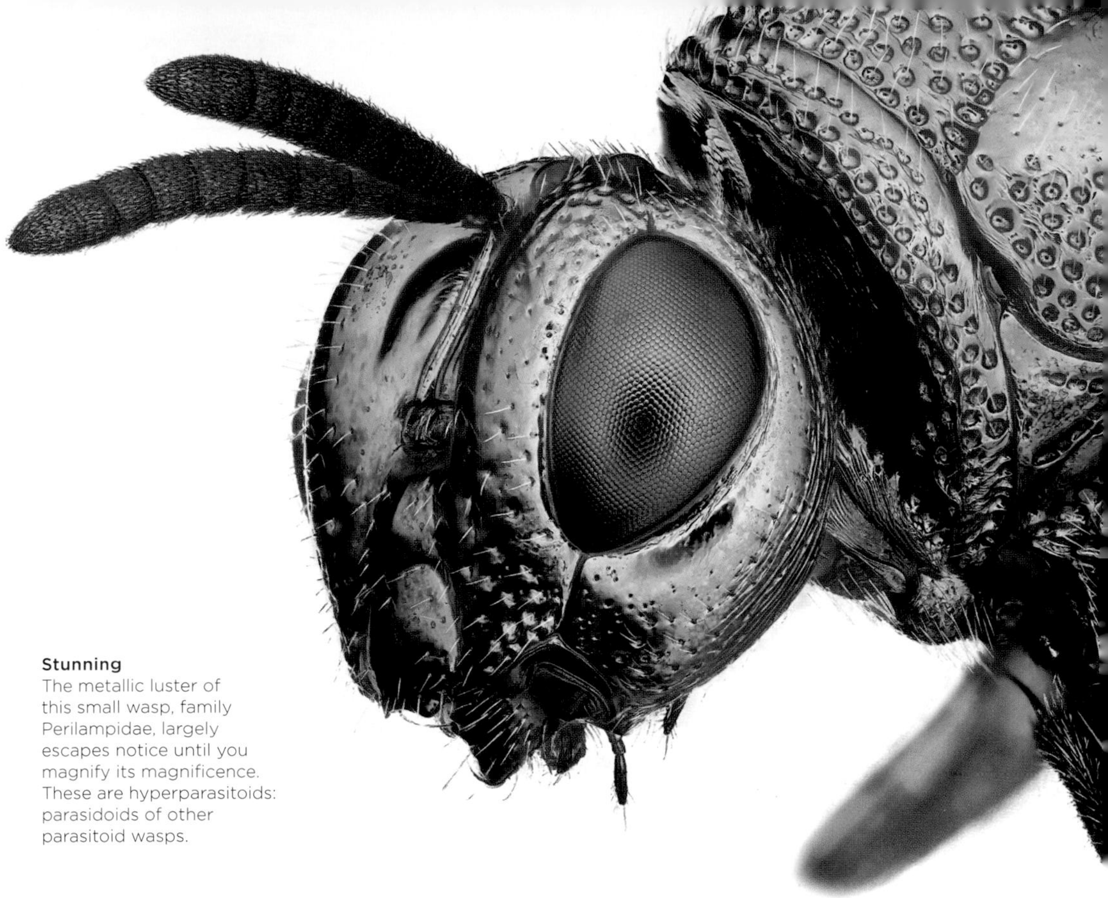

Stunning
The metallic luster of this small wasp, family Perilampidae, largely escapes notice until you magnify its magnificence. These are hyperparasitoids: parasidoids of other parasitoid wasps.

Besides offering wasps the ability to fly, wings can be used in other ways. Wasps may have spots on their wings that give them an uncanny resemblance to an ant, the one insect more likely to be avoided than wasps themselves. Ants seldom travel alone, so predators avoid the prospect of several ants coming to the defense of a nestmate. Flared wings add an intimidating element to threat posturing in wasps. Wing-fanning behavior is an aspect of courtship in some wasp species.

There can be surprising variation within one species of wasp, or within populations, and many species cannot be recognized simply by comparing color patterns to identified specimens. Further complicating matters is that wasps across the entire spectrum of families share the same basic patterns, to reinforce the association of their colors with their habit of stinging. Even male wasps, and those species that do not have stinging females, benefit from this phenomenon.

Stop!
Black and red, as in this *Ammophila* sp. thread-waisted wasp, signal the ability of the insect to sting in self-defense. Combinations of black or metallic blue and yellow, orange, or white, send the same message.

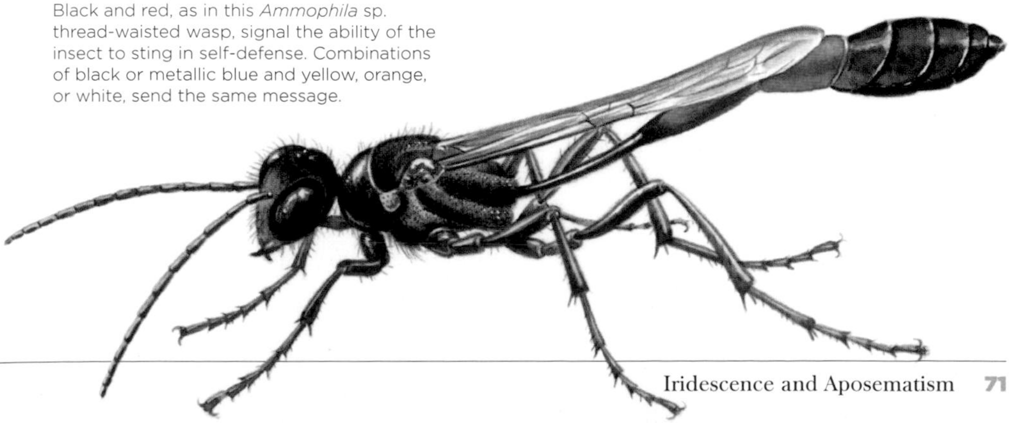

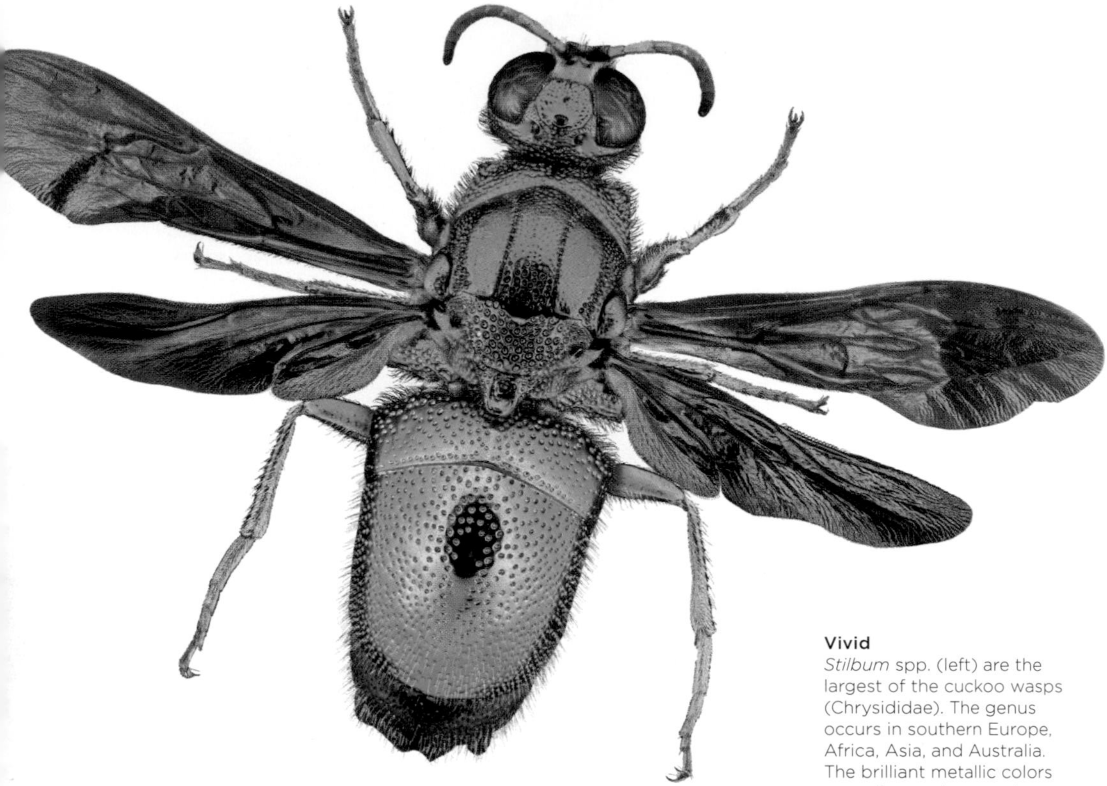

Vivid
Stilbum spp. (left) are the largest of the cuckoo wasps (Chrysididae). The genus occurs in southern Europe, Africa, Asia, and Australia. The brilliant metallic colors come from microscopic structures in the exoskeleton that refract light waves.

Roach Killer
The emerald cockroach wasp (below), *Ampulex compressa*, is large and conspicuous, but still successfully hunts her prey. As in other metallic wasps, her color is a product not of pigment, but of structures that bend light waves.

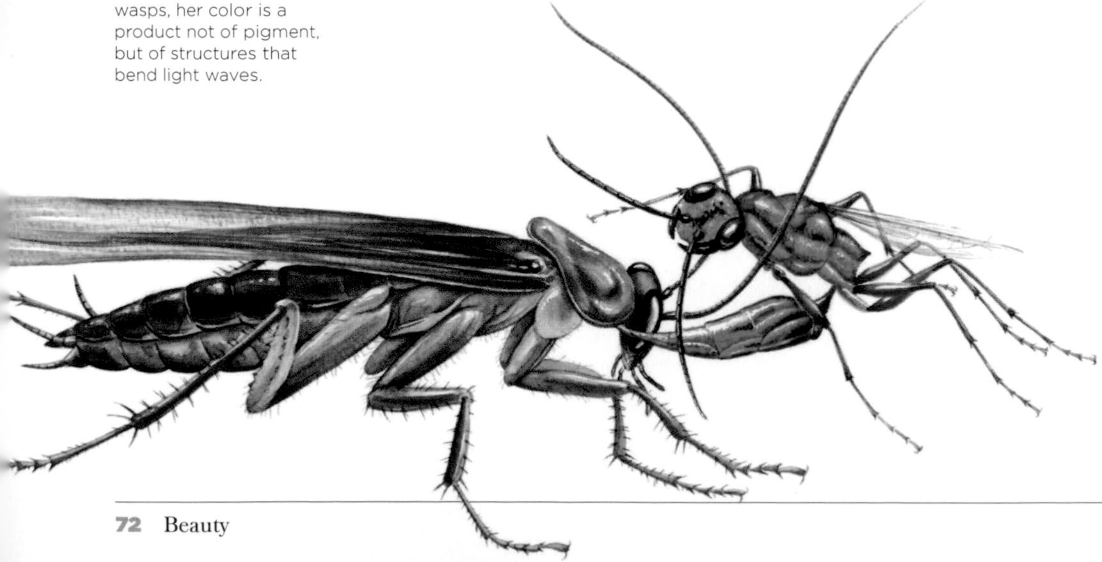

Jewel Wasp
Hedychrum nobile ranges from the UK to southern and eastern Europe. This small jewel is a kleptoparasitoid of certain sand wasps.

Metallic Assassin
The beautiful female *Chlorion lobatum* of southern Asia stings crickets into permanent paralysis, then stores them in an underground burrow as food for her larval offspring.

PROFILE

Cuckoo Wasps

Family Chrysididae

The brilliant reflective colors of many insects are bright enough to make precious stones and metals envious. Gold nuggets, emeralds, and rubies are inanimate after all, whereas cuckoo wasps, torymid wasps, and a number of other wasp families are energetic organisms constantly catching the light as they flit and dash through their surroundings.

Family Chrysididae

SPECIES	~3,000
DISTRIBUTION	Worldwide except for Antarctica
SIZE	0.1–0.7 inches (3–18mm)
AMAZING FACT	Cuckoo wasps do not sting

Actual size

Bright Spot
Despite their small size, cuckoo wasps are highly visible when sunlight strikes their highly reflective bodies. You may only catch a glimpse, though, as they are hyperactive insects.

It may come as a surprise that these vivid metallic colors are not produced by pigments, but are instead generated by fine structures of the cuticle of the exoskeleton. The craters, pits, and reticulations on the bodies of many species of cuckoo wasps (family Chrysididae) have little to do with the greens, blues, violets, reds, and other colors, though the rough topography might make the insects sparkle a bit more. Multiple microscopic layers of the outer cuticle contain microfibrils oriented at various angles that refract certain wavelengths of light, making the insect appear metallic. These are called "interference colors." White light entering the cuticle layers is fractured and bounced, the rays recombining on their exit in ways that express only certain colors with great vibrancy. The angle at which light enters and leaves (angle of incidence) can change the colors we perceive. Consequently, what shows as red at one angle may appear green at a different angle.

The most familiar cuckoo wasps—also known as jewel wasps or ruby wasps—are kleptoparasites in the nests of other solitary wasps and solitary bees. That means that cuckoo larvae feed on the paralyzed insects, spiders, or pollen and nectar stored for the host's larval offspring. As a result of this daring lifestyle, chrysidids have a thick, dense exoskeleton that deflects the bites and stings of host wasps and bees. Cuckoo wasps can also roll into an impregnable ball (conglobulation), much like an armadillo or common pillbug, protecting their heads, antennae, and legs from the gnashing jaws and probing stingers of attacking hosts.

What purpose do these riotous colors serve the wasps? We may never know. It could be that our human eyes are not capable of seeing the entire picture, perceiving all the elements of interference colors. Most insects can see the ultraviolet end of the light spectrum, invisible to us, for example. Insects are also able to make use of polarized light.

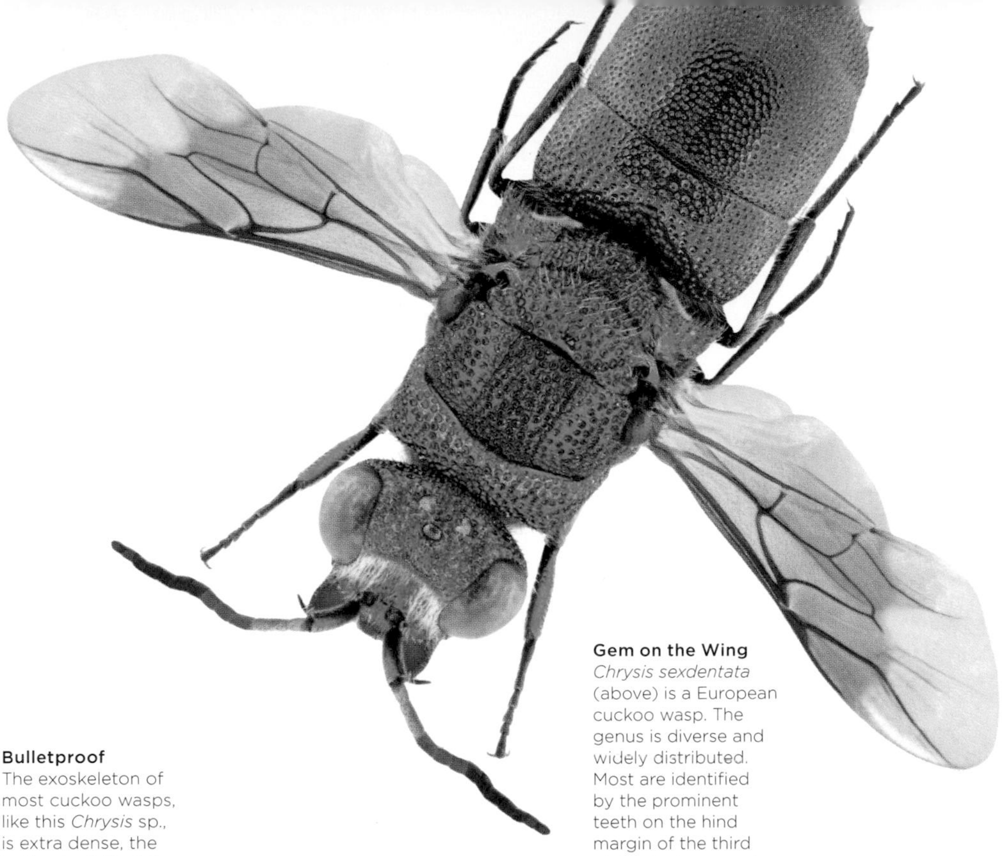

Bulletproof
The exoskeleton of most cuckoo wasps, like this *Chrysis* sp., is extra dense, the better to deflect the bites and stings of its hosts, which are usually the nests of solitary, stinging parasitoid wasps.

Gem on the Wing
Chrysis sexdentata (above) is a European cuckoo wasp. The genus is diverse and widely distributed. Most are identified by the prominent teeth on the hind margin of the third abdominal segment.

Warning Colors

Wasps are often instantly recognized by their patterns of bright colors. Black with horizontal bands of yellow, white, ivory, orange, or red, sometimes with additional spots, vertical stripes, or blotches. This loud ensemble is designed to advertise the ability of wasps to defend themselves with painful, venomous stings. This phenomenon is known as aposematism, from the Greek words for "away" and "sign." It is the opposite tactic to camouflage: make yourself obvious if you are capable of inflicting injury. Inexperienced predators will not forget a painful lesson, and will assume all similarly-colored insects are likewise dangerous.

Mullerian mimicry describes the common wardrobe of warning colors shared by stinging wasp species across many different families. It is a blanket statement that benefits all wasps with the same advertising blueprint. Wasps that congregate on the surface of exposed nests, such as *Polistes*, *Polybia*, *Mischocyttarus*, and other social wasps, or form sleeping aggregations like *Steniolia* sand wasps, *Myzinum* thynnid wasps, and others, derive added benefit from the amplification of warning colors through sheer numbers of individuals clumped together.

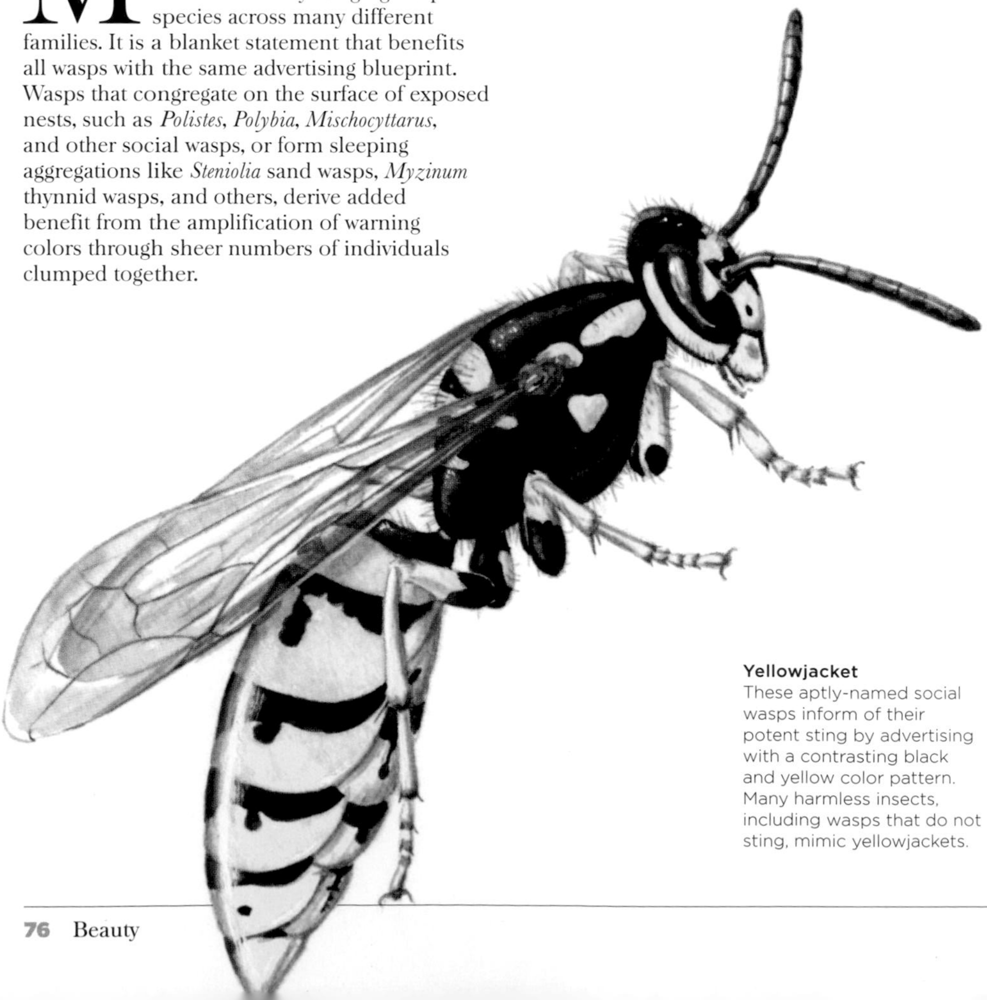

Yellowjacket
These aptly-named social wasps inform of their potent sting by advertising with a contrasting black and yellow color pattern. Many harmless insects, including wasps that do not sting, mimic yellowjackets.

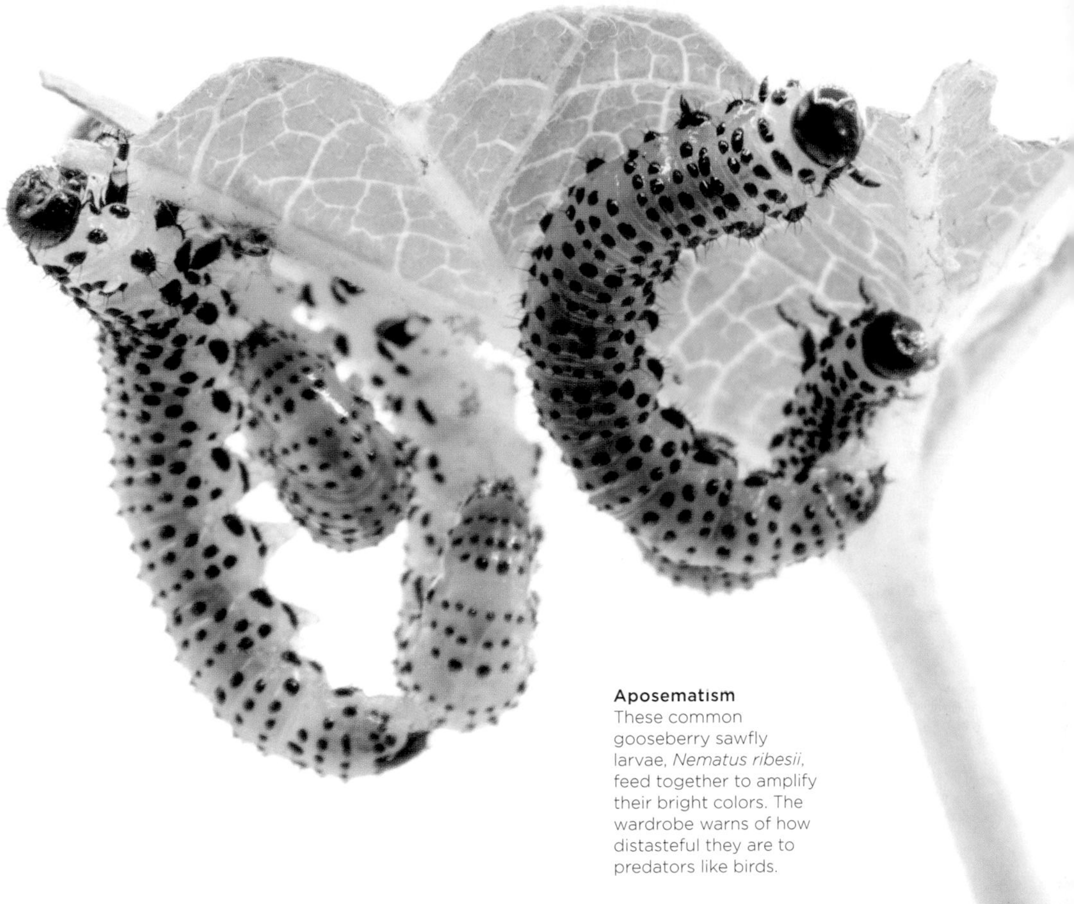

Aposematism
These common gooseberry sawfly larvae, *Nematus ribesii*, feed together to amplify their bright colors. The wardrobe warns of how distasteful they are to predators like birds.

Other behaviors enhance or exaggerate the effect of warning colors. A raised abdomen, flared wings, and tip-toe stance is the classic threat posture of many paper wasps before they launch an attack from their nest. Some wasp species facing imminent danger will curl or contort their abdomens to better expose bright colors normally concealed by their overlapping wings.

The survival strategy of aposematism is so successful that it has been "adopted" via coevolution by perfectly harmless wasps, plus flies, beetles, moths, and other insects. Batesian mimicry represents a type of false advertising designed to dupe potential predators. The masquerade is often complemented by behavioral displays that are stunningly convincing (see Wasp Mimics, page 158).

It is not only adult wasps that exhibit warning colors. Sawfly larvae that feed on toxic plants may sequester those plant compounds in their own bodies, making themselves poisonous to predators. Since the caterpillar-like larvae are often exposed while feeding, they may be brightly colored to advertise their potency. Many sawfly larvae also feed gregariously, broadcasting their "don't eat me" message in a more effective manner. Defensive compounds can be regurgitated from special storage sacs in the foregut, the diverticulae, or the hindgut (anus), depending on the species, when the larva rears back in a threat posture.

Warning colors are almost always pigment. Pigments work by absorbing all wavelengths of light except for the one they express. Yellow represents a pigment that subtracts all wavelengths in the spectrum except yellow, for example. Black results from an absorption of the entire white light spectrum. White results from a reflection of incoming white light.

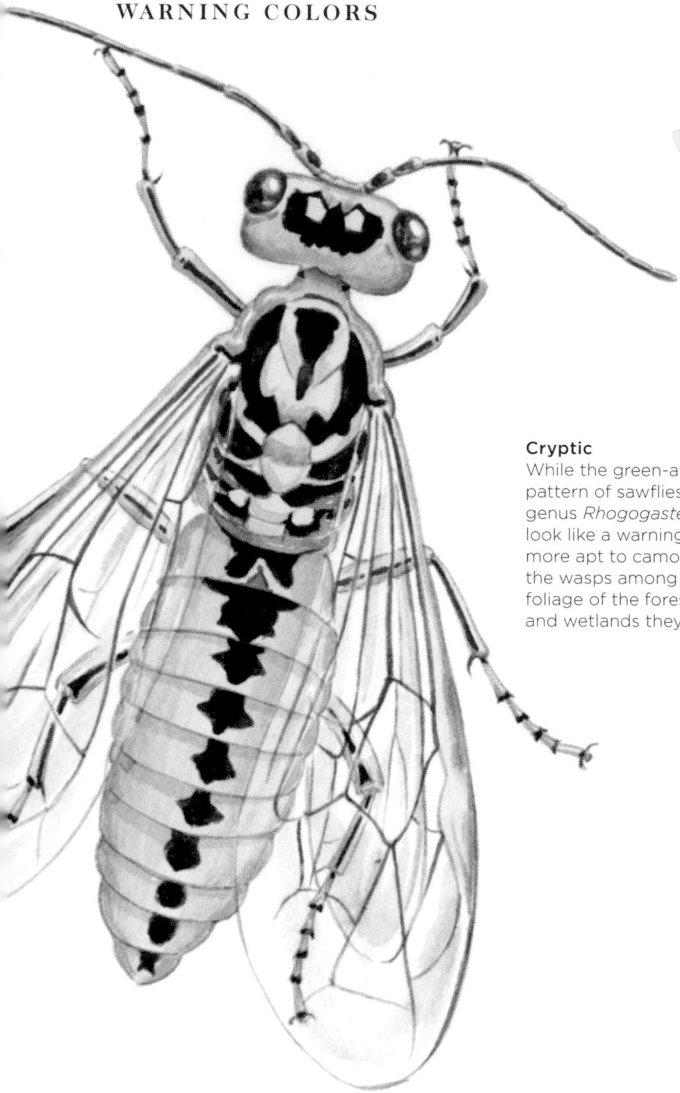

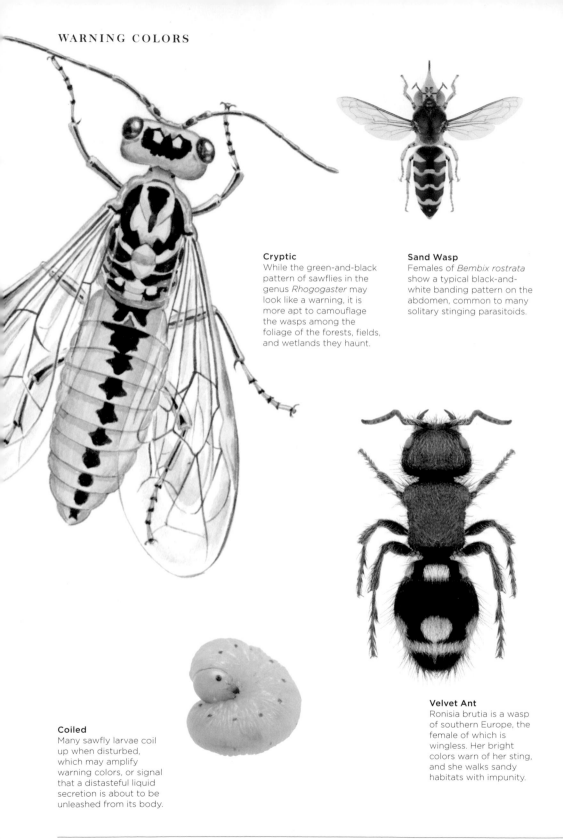

Cryptic
While the green-and-black pattern of sawflies in the genus *Rhogogaster* may look like a warning, it is more apt to camouflage the wasps among the foliage of the forests, fields, and wetlands they haunt.

Sand Wasp
Females of *Bembix rostrata* show a typical black-and-white banding pattern on the abdomen, common to many solitary stinging parasitoids.

Velvet Ant
Ronisia brutia is a wasp of southern Europe, the female of which is wingless. Her bright colors warn of her sting, and she walks sandy habitats with impunity.

Coiled
Many sawfly larvae coil up when disturbed, which may amplify warning colors, or signal that a distasteful liquid secretion is about to be unleashed from its body.

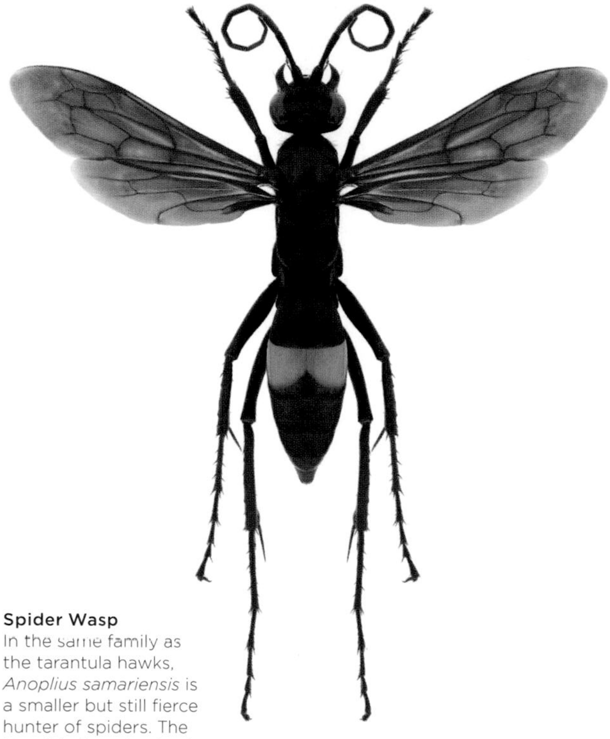

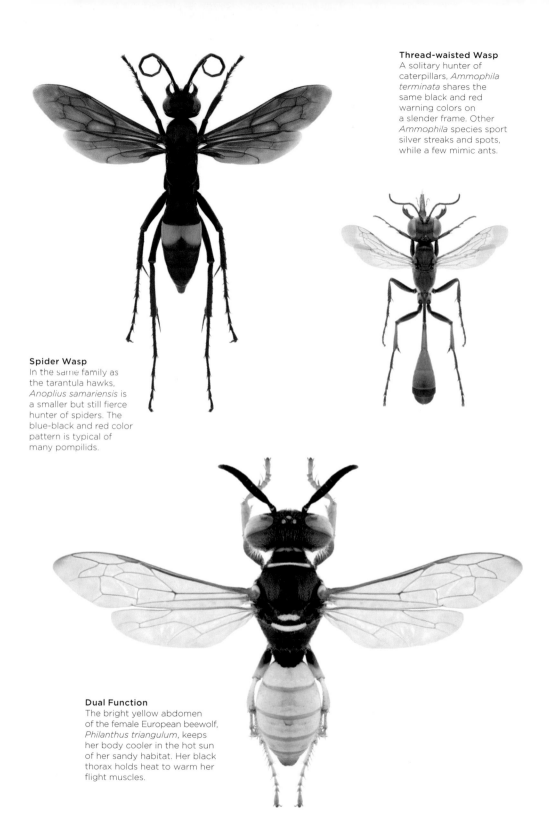

Thread-waisted Wasp
A solitary hunter of caterpillars, *Ammophila terminata* shares the same black and red warning colors on a slender frame. Other *Ammophila* species sport silver streaks and spots, while a few mimic ants.

Spider Wasp
In the same family as the tarantula hawks, *Anoplius samariensis* is a smaller but still fierce hunter of spiders. The blue-black and red color pattern is typical of many pompilids.

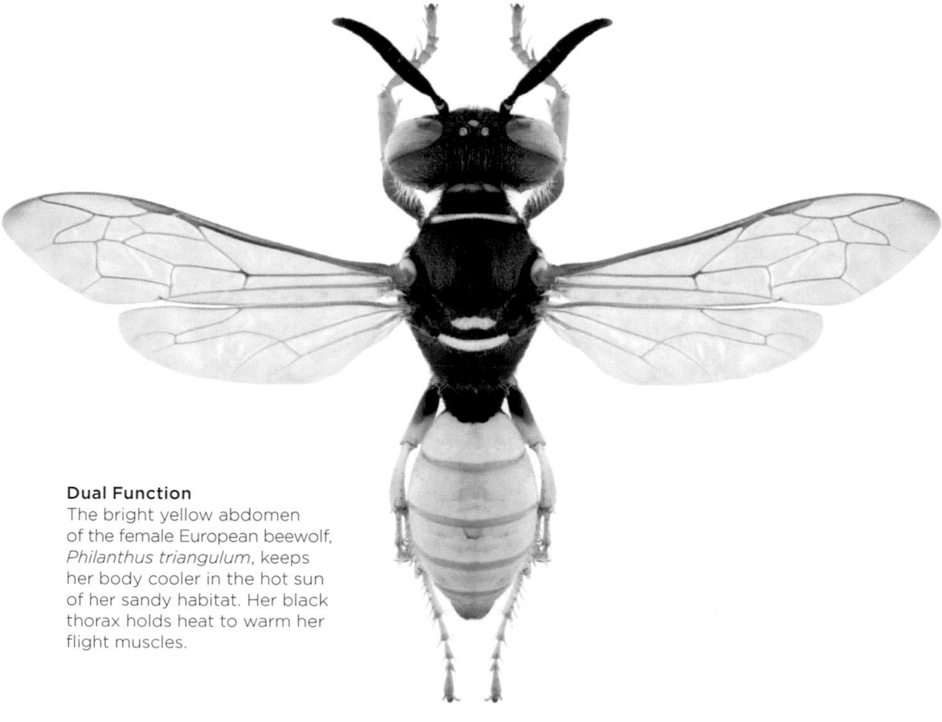

Dual Function
The bright yellow abdomen of the female European beewolf, *Philanthus triangulum*, keeps her body cooler in the hot sun of her sandy habitat. Her black thorax holds heat to warm her flight muscles.

PROFILE

Tarantula Hawks

Genera *Pepsis* and *Hemipepsis*

A different combination of warning colors is demonstrated by some members of the spider wasp family Pompilidae. The tarantula hawk, *Pepsis heros*, is probably the largest stinging wasp, as shown below at actual size. Found in the northwestern third of South America, it models the iridescent blue body and bright orange or scarlet wings common to many large spider wasps. These colors are structural in origin, restricted to appressed pubescence (fine, densely matted, scale-like hairs) on the wasp's body. The wings are pigmented, but may also be covered in iridescent scales.

Actual size

Family Pompilidae

SPECIES	~135 *Pepsis*, ~132 *Hemipepsis*
DISTRIBUTION	Worldwide, mostly deserts and tropics
SIZE	2 inches+ (50mm+)
AMAZING FACT	*Pepsis grossa* is the official state insect of New Mexico, U.S.A.

Giant Beauty
Pepsis grossa ranges from the southern U.S.A. to northern South America. Some populations have individuals with black wings. In South America a "lygamorphic" variety has dark wings with amber patches and pale tips.

A Flashy Wardrobe
The bright orange wings of this *Pepsis* sp. from Argentina stand in stark contrast to her steel blue body, advertising to potential predators her ability to deliver a painful sting.

At least some tarantula hawk species reinforce their warning colors with aromatic compounds secreted by the mandibular glands near their jaws, and threat posturing with abdomen curled under and forward, and wings splayed. The wasps also have a dense exoskeleton that helps repel attacks from the fangs of their spider prey, the beaks of birds, and jaws of other adversaries.

Pepsis species found in scorching desert habitats often assemble in loose, usually active aggregations in the shade provided by the foliage of trees or shrubs to pass the hottest hours of the day. Both sexes are usually present, so the gatherings might also function as the insect equivalent of speed dating. The communal clusters also increase the impact of their warning colors when so many individuals are on display at once.

Ignore the warnings at your peril. The Schmidt Sting Pain Index ranks *Pepsis* stings a whopping four out of four: "Blinding, fierce, shockingly electric. A running hair dryer has just been dropped into your bubble bath." Surprisingly, the effects of a tarantula hawk's self-defense sting begin and end there. A human with a healthy immune system typically experiences immediate, excruciating pain from a *Pepsis* sting, but the agony subsides quickly, in roughly three minutes or so. There is no nerve damage, tissue damage, or any other sub-lethal or chronically debilitating effect of the venom.

The tarantula prey of these wasps does not get off so easily, though, as her venom induces permanent paralysis. The victim is then dragged to a pre-existing cavity, sometimes the spider's own burrow, where the limp arachnid is deposited. Our spider-slaying heroine places a single egg on the still-living host. The larva that hatches will consume the spider.

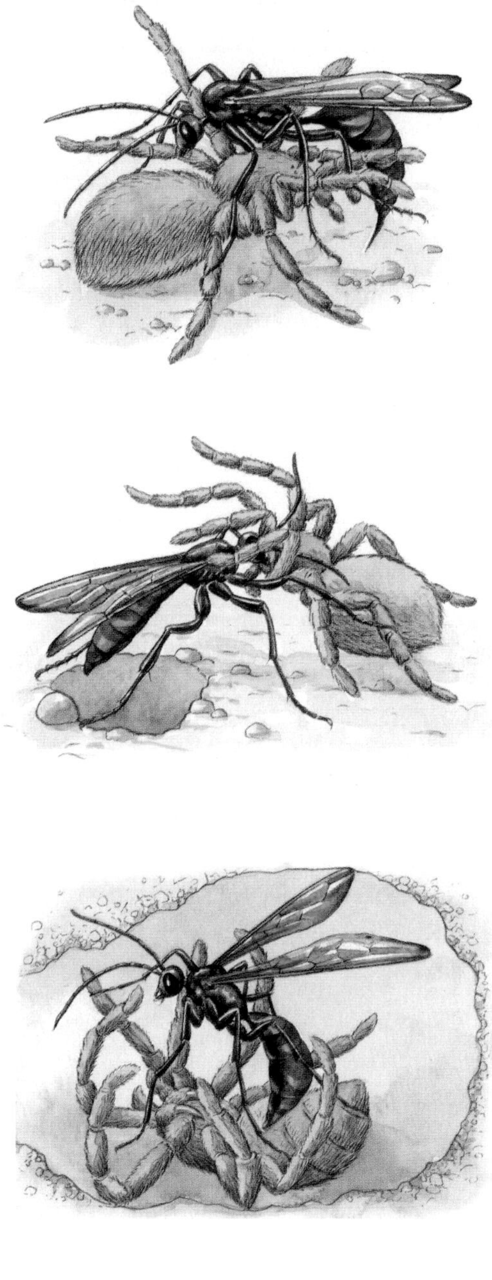

An Epic Wasp Versus Tarantula Battle
A tarantula hawk from New Mexico, U.S.A., stings a spider into paralysis, then drags it into a burrow. She lays an egg upon it, and the larva that hatches feasts.

PROFILE

Velvet Ants

Family Mutillidae

It may be tempting to reach down and stroke the cute, furry "ant" crossing your path, but don't. Velvet ants are the wingless females of solitary wasps in the family Mutillidae, and they pack an excruciating sting. The most conspicuous species are covered in a dense coat of long, bright orange, yellow, or red hairs as a warning of their hidden weaponry. Those that are less setose fool potential predators into mistaking them for true social ants that can sting, spray formic acid, or quickly summon large numbers of their sisters if they are attacked. Some species in Australia and neighboring northern islands are metallic bronze, green, blue, or violet.

"Cow Killer"
The female Eastern Velvet Ant, *Dasymutilla occidentalis* of North America (left), has a sting so painful it can supposedly kill livestock. Pure myth, thankfully.

Male velvet ants are usually fully winged and often much larger in size than their female counterparts. Scientists have yet to associate all males and females, so some species are known by only one sex.

Velvet ants can be diurnal, but those that inhabit hot, arid, open habitats tend to be active on overcast days, are crepuscular, or even nocturnal. These wasps are parasitic on other insects, so females spend most of their time seeking hosts. Males spend their time seeking mates. Our picture of host relationships for velvet ants is as fuzzy as the insects themselves, but we know that many are, as larvae, parasitic on the mature larvae or pupae of other solitary wasps or solitary bees. Consequently, female mutillids have perhaps the thickest exoskeletons of any hymenopterans, the better to protect them from stinging, biting hosts.

A surprising aspect of velvet ant behavior is the ability of both sexes to produce sound by stridulation—the rubbing together of two body parts. The second and third dorsal segments of the abdomen (tergites) act in concert to produce an audible squeak. A "scraper" on the ventral edge of tergite two rasps across a series of parallel ridges on the upper surface of the third tergite to make the sound. The frequency of sound produced appears to be species-specific. This is employed as a "distress call" when the insect perceives danger, but in at least one species stridulation is part of courtship and mating, too. A male of *Dasymutilla foxi* will pounce on a female, vibrating his wing muscles to produce a "honk," to which the female responds with "chirping" stridulations. Once joined in copulation, the male ceases honking and joins the female in stridulating for the duration of their engagement.

Family Mutillidae

SPECIES	~5,000
DISTRIBUTION	Worldwide except for Antarctica
SIZE	0.1–1 inches (3–25mm)
AMAZING FACT	Longest stingers of any hymenopteran

Actual size

Rock Crawler
Females of the velvet ant *Physetopoda halensis* inhabit rocky steppe habitats, and sometimes sandy areas in Europe. The hosts for this parasitoid wasp remain unknown.

5

Ecology
The Place of Wasps in Ecosystems

Wasps Make the World Go Round

While some entomologists study wasps at the level of molecular DNA, other scientists decipher the place of wasps in the bigger picture of niches, habitats, and the biosphere. Wasps fill almost every imaginable role in ecosystems. Their impact on humanity may be direct or indirect, but all species are indispensable in the grand scheme.

Insects in general provide human benefits under the umbrella of "ecosystem services." This includes pollination services to crops and wildflowers, and food for other wildlife we enjoy through observation, hunting, or angling. Recycling of animal waste and decaying organic matter is yet another function. Insects control pests in agricultural and forest ecosystems, too. Wasps figure into almost all of those categories. A study published in 2006 estimated the value for ecosystem services provided by insects in the U.S.A. alone is more than 57 billion U.S. dollars annually.

Wasps are obligatory pollinators of a surprising number of plants, from orchids to figs. Wasps are part of the food web in this regard as they help propagate the plants that feed other animals. Sawflies, gall wasps, and other wasps are primary consumers of plants, and are sometimes crop, garden, or forest pests from the human perspective. Those same species are food for other animals, including many other wasp species that are parasitoids or predators of eggs, larvae, or adults. We employ some as biocontrols to alleviate competition for what we view as "our" resources. Meanwhile, horntail wasps and their kin rejuvenate forests by helping decompose dead wood. They are in turn food for the giant ichneumon wasps that are their enemies.

To accomplish their jobs in ecosystem services, wasps may engage in mutualistic relationships with completely different organisms. Certain braconid wasps employ a virus to assist in disabling the immune system of the wasp's host. Female horntail wasps deposit a special fungus along with their eggs so that their larval offspring will have help in digesting the cellulose in their diet. Fig wasps and their host plant are yet another example of symbiosis.

We like to think we are insulated from natural processes and other organisms, but wasps can even play out their lives inside our homes. They use buildings as shelters for their own nests, raid our picnics for food, eat roses and other garden plants, and hover around lights at night. We compound matters by accidentally or intentionally importing wasps from foreign countries that then become invasive species. Slowly, however, our tolerance for wasps is growing, and we even breed them for release into croplands where they control pests. Continuing to harness the power of these eco-warriors will benefit us *and* wasps.

Same Target?
Two female giant ichneumon wasps, *Megarhyssa macrurus*, may have detected the same host: a grub of the pigeon tremex horntail, *Tremex columba*, boring inside a log.

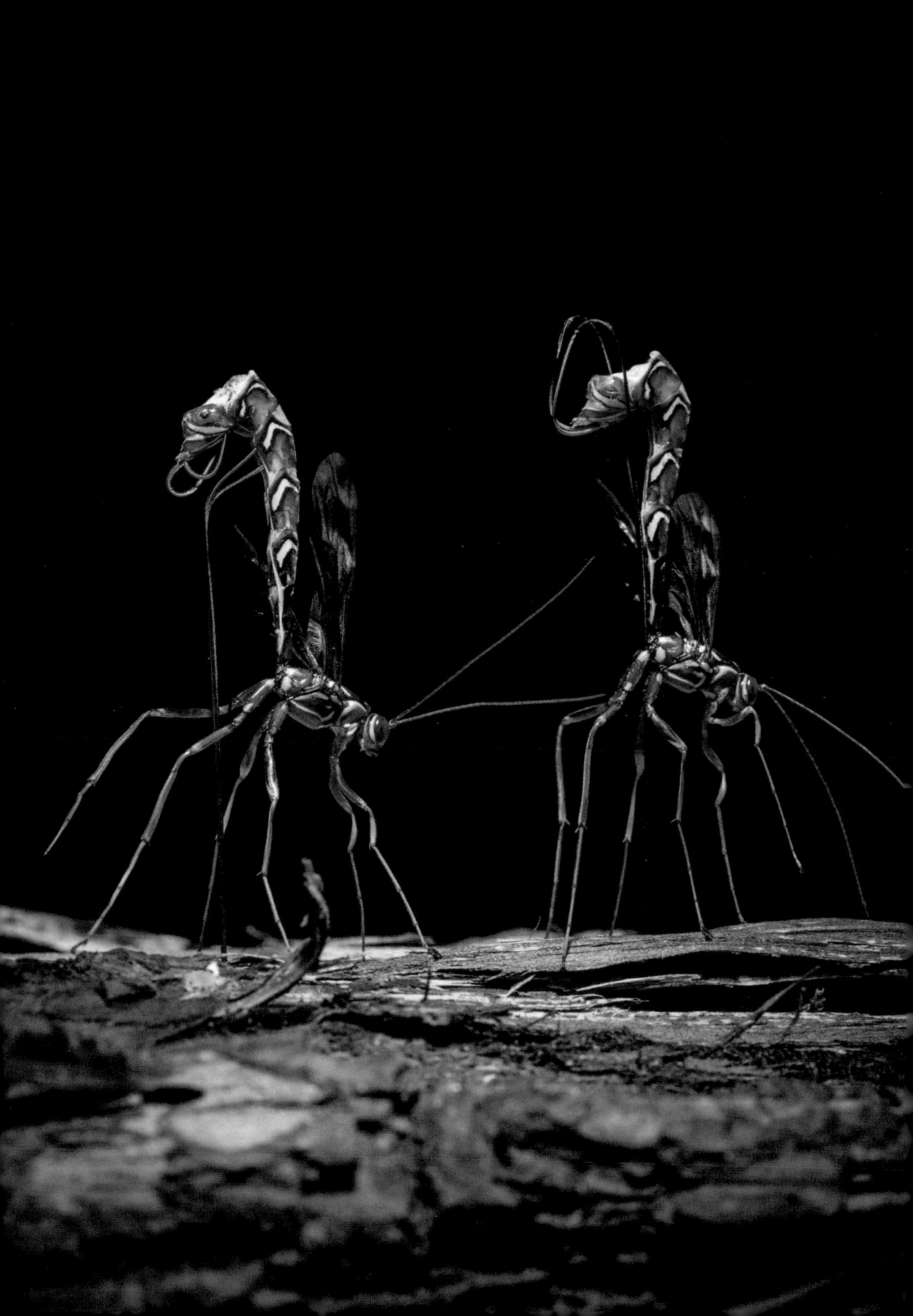

Pollinators

Bees are regarded as the primary pollinators of flowers, but wasps are the exclusive agents of pollination for a surprising number of plants, especially some orchids. This relationship often involves outrageous sexual deception on the part of the flower, luring only male wasps seeking virgin females. After all, if it smells like a wasp, looks like a wasp, and feels like a wasp…

Hairy Giant
A female Mammoth Wasp, *Megascolia maculata flavifrons* of Europe, seeks nectar in a flower. Scoliid wasps are hairier than most wasps and so accomplish pollination more effectively, though still unintentionally.

The classic example of wasp-mimicking flowers is the mirror orchid (aka "looking-glass orchid"), *Ophrys vernixia*, of Mediterranean Europe and northern Africa. It bears an impressionistic resemblance to a female scoliid wasp, *Dasyscolia ciliata*. The flower's fragrance parallels the sexual scent of the female wasp to get a male's attention. The shiny blue of the flower, reflecting ultraviolet light, is an even better beacon since it mimics the iridescent blue wings of the lady wasp.

Upon landing, the male receives tactile stimuli from the flower that also evokes a female wasp. He proceeds to attempt mating. Not surprisingly, this floral deception is called "pseudocopulation."

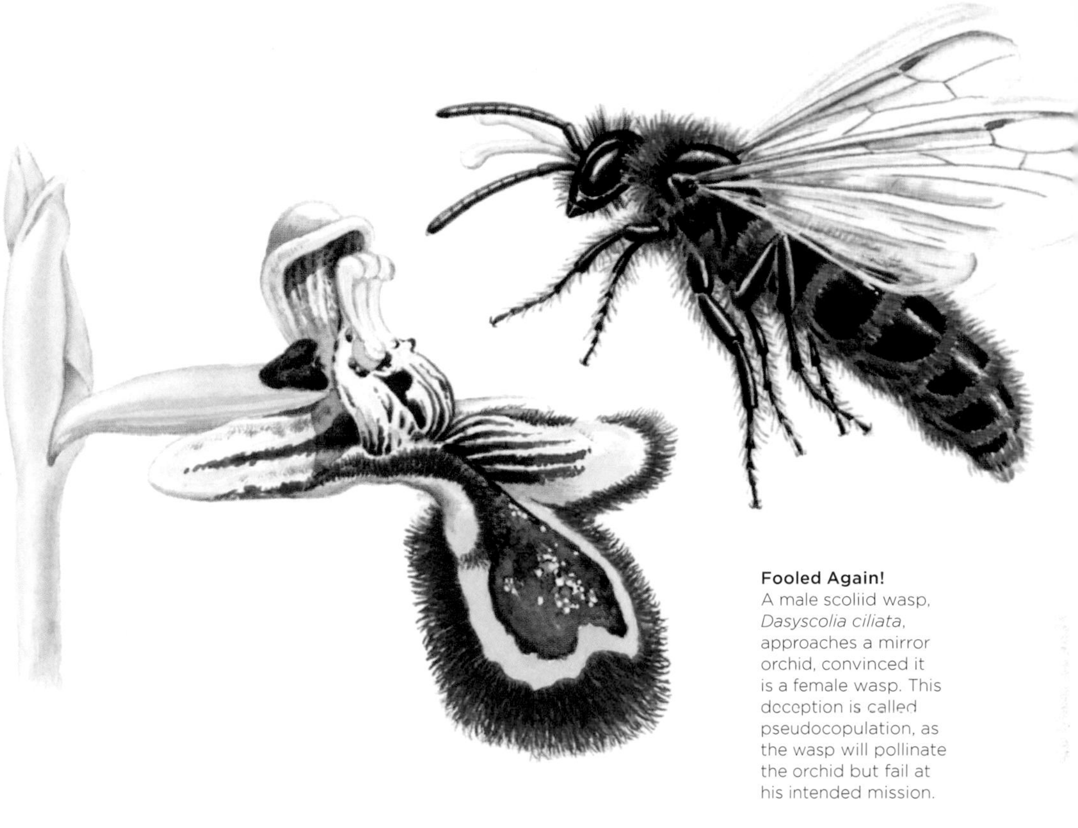

Fooled Again!
A male scoliid wasp, *Dasyscolia ciliata*, approaches a mirror orchid, convinced it is a female wasp. This deception is called pseudocopulation, as the wasp will pollinate the orchid but fail at his intended mission.

Pseudocopulation is practiced by many other orchids, from Japan to South America. Eleven genera of orchids in Australia alone feature this strategy. The tongue orchids *Cryptostylis erecta* and *C. leptochila* are pollinated solely by a male ichneumon wasp, *Lissopimpla excelsa*. The deceit is so complete that the male may waste a load of sperm on the phony female. Sexual deception of *Ophrys subinsectifera* fools the argid sawfly *Sterictophila gastrica* in northern Spain, while the Australian orchid *Caleana major* benefits from naïve male pergid sawflies in the genus *Lophyrotoma*.

African orchids *Eucomis* spp, *Disa sankeyi*, and *Satyrium microrrhynchum* are pollinated mostly by tarantula hawks in the genus *Hemipepsis*. The African milkweed *Pachycarpus asperifolius* is likewise pollinated by these giant wasps. Orchids and milkweeds both deliver pollen in packets that adhere stubbornly to the insects until they are delivered to another plant. The packets, called pollinia, each fit like a key in the lock of another flower.

South Africa and Australia are epicenters of the Masarinae or "pollen wasps," a subset of the Vespidae that includes yellowjackets, hornets, and mason wasps. The Australian masarine *Rolandia angulata* is likely the principal pollinator of *Goodenia* flowers. In the southwest U.S.A., *Pseudomasaris vespoides* is a nearly exclusive pollinator of violet-colored *Penstemon* flowers. Pollen wasps harvest pollen as food for their offspring, in contrast to the customary meat-based diet of most larval wasps. The only other wasps known to provision with pollen are square-headed wasps in the genus *Krombeinictus*.

Of all the examples of co-evolution between flowers and wasps, none are as complicated and bizarre as the fig wasps. Turn this page to find out why.

PROFILE

Fig Wasps

Family Agaonidae

Figs, from the *Ficus* in the office lobby to the fruit in the grocery, are familiar and diverse. Fig fruits account for up to 70 percent of the diet of some tropical rainforest vertebrates, as well as being an important cash crop. Most fig species rely on mutualistic relationships with miniscule wasps in the family Agaonidae. More than half of the nearly 800 agaonids are parasitoids of other insects and, in reality, probably belong in other wasp families. That leaves at least 368 species that are fig pollinators.

Family Agaonidae

SPECIES	640 described ("known"), probably >1,300
DISTRIBUTION	Tropical: Americas, Africa, Asia, and Indo-Australian regions
SIZE	<0.08–average is about 0.06 inches (<2mm–average is about 1.5mm)
AMAZING FACT	The only creatures that pollinate figs

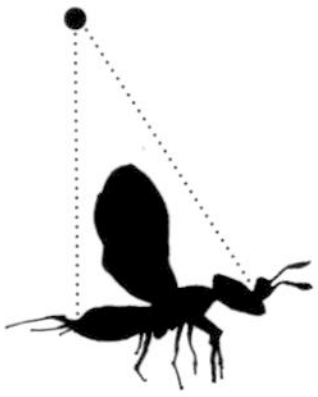

Tiny Heroine
Female fig wasps are less than two millimeters in size, but the role they play in tropical ecosystems is infinitely larger. Many other animals depend on ripe figs for nutrition and energy.

You are unlikely to ever see a fig flower. They are tiny, concealed within a fruit-like synconium. A female wasp is attracted to a synconium by scents from the flowers inside that are broadcast through a small hole called the ostiole. She enters through the ostiole, and may lose body parts squeezing through. Once inside the cavernous synconium, she begins attempting to lay eggs, one in each female flower lining the interior. It is this act that accomplishes pollination. Her ovipositor cannot, however, reach the ovary of every flower. These "taller" flowers, free of wasp eggs, will become normal, successful seeds. In cases where she lays her egg, her larva offspring lives, feeds, and grows within the developing fig seed, eventually emerging as an adult wasp. Only a fraction of fig seeds is destroyed, the remainder left to be dispersed by vertebrate animals and eventually becoming new fig trees.

While female fig wasps look like delicate fairies, the males resemble grotesque dragons, wingless or nearly so, wormlike, essentially blind, with beefy legs and menacing jaws. Males may fight each other in competition over emerging females. Mating takes place within the darkness of the synconium, before the female wasps bore their way through the walls to freedom. Prior to their exit, the newly-mated females become dusted with pollen from male flowers inside the synconium.

Non-pollinator
Many wasps associated with figs, like these female pteromalids, *Apocrypta* sp. (below), make unauthorized entry from the exterior. They are usually parasitoids of fig wasps or other parasitoids, or inquilines ("guests") that do no harm.

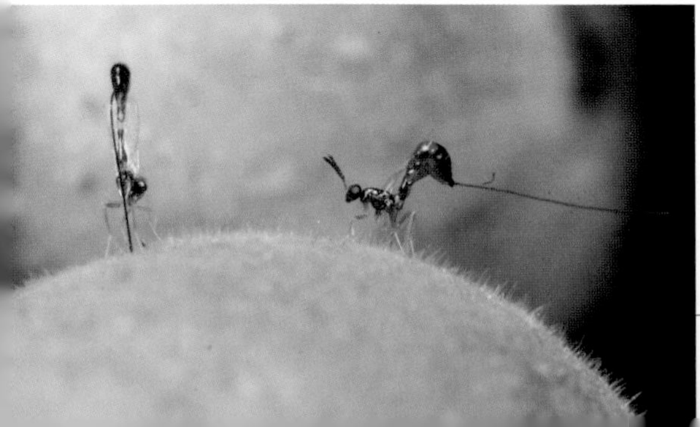

Fig Pollination
Please see text for a full explanation of the cycle. Fig flowers are hidden inside the fruit-like synconium, and can only be pollinated by miniscule wasps that have co-evolved to do so.

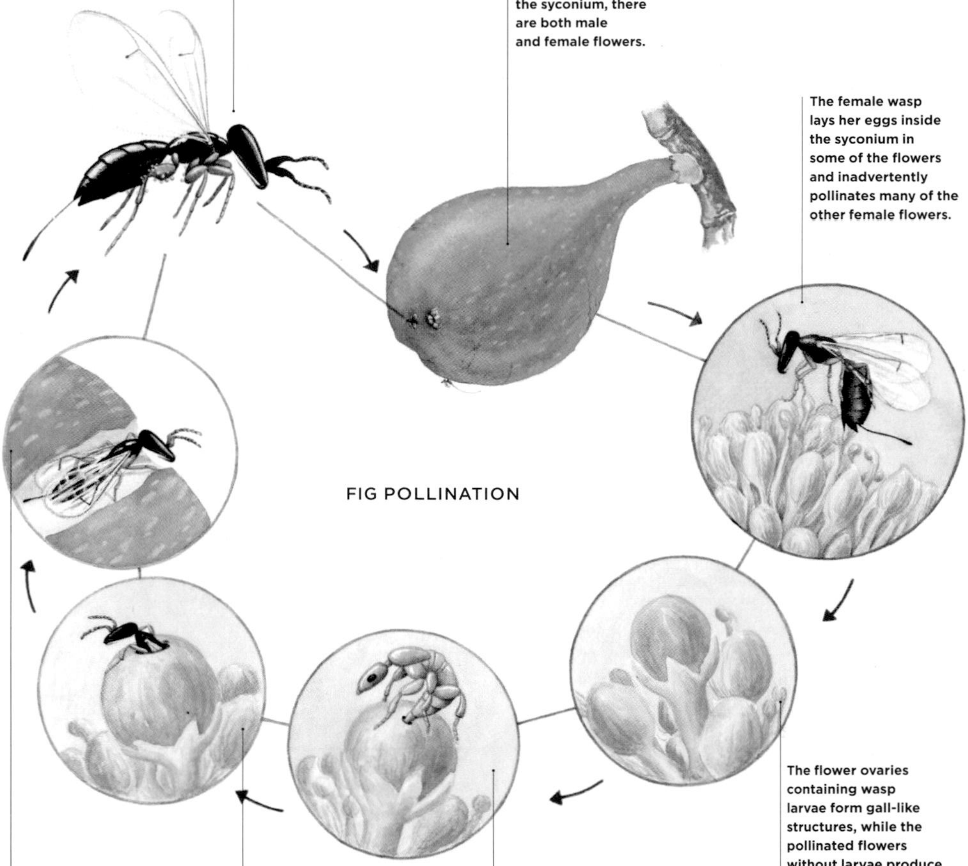

The pollen-laden female wasp enters the syconium of the unripe fig through an opening called the ostiole.
Inside the fig, within the syconium, there are both male and female flowers.
The female wasp lays her eggs inside the syconium in some of the flowers and inadvertently pollinates many of the other female flowers.
FIG POLLINATION
The flower ovaries containing wasp larvae form gall-like structures, while the pollinated flowers without larvae produce seeds instead.
The male wasps are the first to emerge from the galls as the fig ripens. They roam around the syconium looking for female wasps to fertilize while still contained in their galls. The wingless male wasps will never leave the fig. Their final task is to dig escape routes for the female wasps and then they die.
The mated female wasps break free of their galls just as the male flowers within the syconium reach maturity.
The female wasps collect pollen en route to the escape tunnels created by the male wasps. However, some females will exit the fig without collecting pollen. Regardless, the female wasp will now go in search of another unripe fig in which to lay her eggs.

The life cycle described here is generic and oversimplified, but male fig wasps never leave the fig they are born in, and females die within the synconium after laying their eggs. Some figs produce male flowers on one tree and females on another. Some fruits are inedible galls. Some figs need no pollination, and those include many common landscape varieties.

The Food Web

Scarcely any trophic level in the food web is absent of at least one wasp species. Most wasps fill the niches of parasitoid or predator of some other invertebrate animal, but as we are beginning to see, many wasps have completely different roles. Some are pollinators, some scavengers, some vegetarians. All can be food for other organisms, completing the meandering cycle that is the flow of energy, the overriding currency of ecosystems.

Plants are the foundation of terrestrial ecosystems. They convert solar energy, water, and soil nutrients into consumable calories through photosynthesis, but it takes pollinating insects—wasps included—to ensure that plants reproduce themselves. Primary consumers come in the form of vegetarian or omnivorous organisms, and wSasp larvae are among those feasting on leaves, stems, buds, flowers, seeds, and fruits. Plant galls, abnormal growths stimulated by insects, mites, fungi, bacteria, and other organisms, can be entire ecosystems unto themselves. What goes into an insect-generated gall is seldom what comes out. Innumerable parasites, predators, and inquilines (squatters) plague the original gall-maker, and wasps are a majority of them.

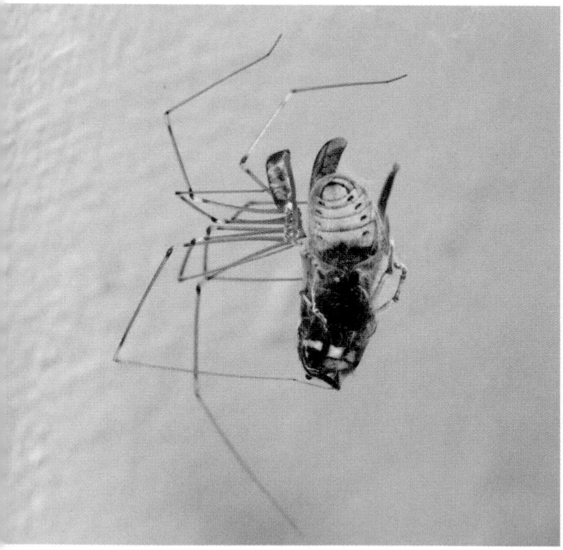

Females of wood-boring wasps, like horntails and their kin, habitually select weakened, dying, or newly dead trees as hosts for their wood-boring larvae. The grubs, and the associated fungi they carry with them, hasten the decomposition of the wood, helping return nutrients to the soil where younger, healthier generations of trees can prosper.

Some plants turn the tables on wasps and other insects. One strategy for overcoming nutrient-poor soils, especially those lacking nitrogen and phosphorus, is to become a bug trap. The false promise of nectar lures wasps and other insects into the wells of pitcher plants, the tentacles of sundews, or the jaws of a Venus flytrap.

Many animals prey on wasps, including mammals that tear open yellowjacket nests and endure multiple stings to feast on the helpless larvae and pupae. Meanwhile, birds pick off wasps in mid-air and bash them against branches to disarm the sting, while spiders entangle wasps in their silken snares. *Cordyceps* fungi infiltrate the bodies of wasps, forcing the insect to become an incubator and vessel for the dispersal of spores to still other victims. The bodies of deceased wasps, however they meet their fate, are ultimately recycled by bacteria and other organisms, completing the cycle by furnishing plants with additional soil nutrients.

Silken Trap
A cellar spider, family Pholcidae, has captured a yellowjacket in her web. Most spiders that kill wasps are species that do not spin webs, but wait in ambush on flowers instead.

The complexity of food webs cannot be understated. Unless a given insect is of economic importance to humanity, (either positive or negative), we devote precious few financial resources to learning about it. Consequently, our collective knowledge of wasps is sorely lacking in many respects, especially in the ecological sense. Your own observations, carefully documented, could reveal previously unknown relationships.

Entangled
Seeking liquid sustenance, a small wasp succumbs to the clutches of a sundew plant, *Drosera capensis*, in South Africa.

Oh, Snap!
The European bee-eater, *Merops apiaster*, is one of many birds that routinely prey on wasps. Birds are quick enough to snatch the insects in mid-air, and their horny beaks deflect stings.

Mutualistic Relationships

As described on the previous page, fig wasps and fig trees are an example of symbiosis called "mutualism," where both species benefit from the relationship. In other wasps, mutualism includes mites, fungi, viruses, or bacteria. The assemblage of a wasp and its microbial associates is itself an ecosystem called a holobiont. You and the bacteria in your gut are also a holobiont.

Zombie Caterpillar
This moth larva serves as an involuntary bodyguard for the cocoons of wasp larvae (Braconidae: Microgastrinae) that once fed inside it, thanks to a virus that the mother wasp injected along with her eggs.

Larvae of woodwasps in the families Siricidae (horntails) and Xiphydriidae require fungal symbionts supplied by the mother wasp in order to make wood digestible. The adult female wasp has special reservoirs called mycangia, located at the base of her ovipositor, that contain fungal spores. She deposits spores, together with mucus manufactured in abdominal glands, when she lays an egg inside a tree. Each fungus may or may not be specific to one species of wasp, but the fungi depend on the wasps for transport.

Fungi can be an enemy, too, so some wasps sanitize the cribs of their offspring. The female European beewolf, *Philanthus triangulum*, cultivates a *Streptomyces* bacteria in specialized antennal glands, which she applies to the walls of her underground nest cells. The larva that hatches from each egg takes up the bacteria as protection, and even impregnates the cocoon it spins before pupating.

Females of some mason wasps, solitary members of the family Vespidae, have special chambers called acarinaria that serve as "carports" for mites. The acarinaria are pockets located at the front of the abdomen. The mites were assumed to be parasitic until it was discovered that in at least one species, *Allodynerus delphinalis*, its symbiotic mite acts as a warrior. Pupae and prepupae of the mason wasp are vulnerable to the parasitoid wasp *Melittobia acasta*, but the mites from the mason wasp disembark to gang up on the female *Melittobia* wasps.

Stranger still is the case of polydnaviruses (PDVs), peculiar to wasps in the families Braconidae and Ichneumonidae. About 18,000 species of braconids are known to possess bracoviruses. Both braconids and ichneumons (ichnoviruses are their associates) replicate polydnaviruses in the calyx region of the female wasp's ovaries. The PDVs are passed on in the egg stage, injected into the wasp's host, such as a moth caterpillar. The virus does not replicate inside the host, but does act to suppress its immune system, halt its growth, or even alter its behavior. The host is left defenseless against its internal parasitoid. Oddly, the venom of the mother wasp may be necessary to activate the virus. The freakish results include "brainwashed caterpillars" that guard the cocoons of their "voodoo wasp" (*Glyptapanteles* spp.) parasitoids, thrashing involuntarily in the presence of a potential enemy of the wasp cocoons.

Voodoo Wasp
The life cycle of *Glyptapanteles* sp. begins when the female lays eggs (yellow oblongs) in the host, along with a virus (blue dots) that accumulate in her calyx. The wasp larva feeds internally while the virus takes over the caterpillar's nervous and immune systems. Eventually, the wasp larva pupates and the virus begins to reproduce there. Eventually an adult wasp emerges from the pupa. Approximately 18,000 species of wasps are associated with bracoviruses. These viruses enable successful parasitism.

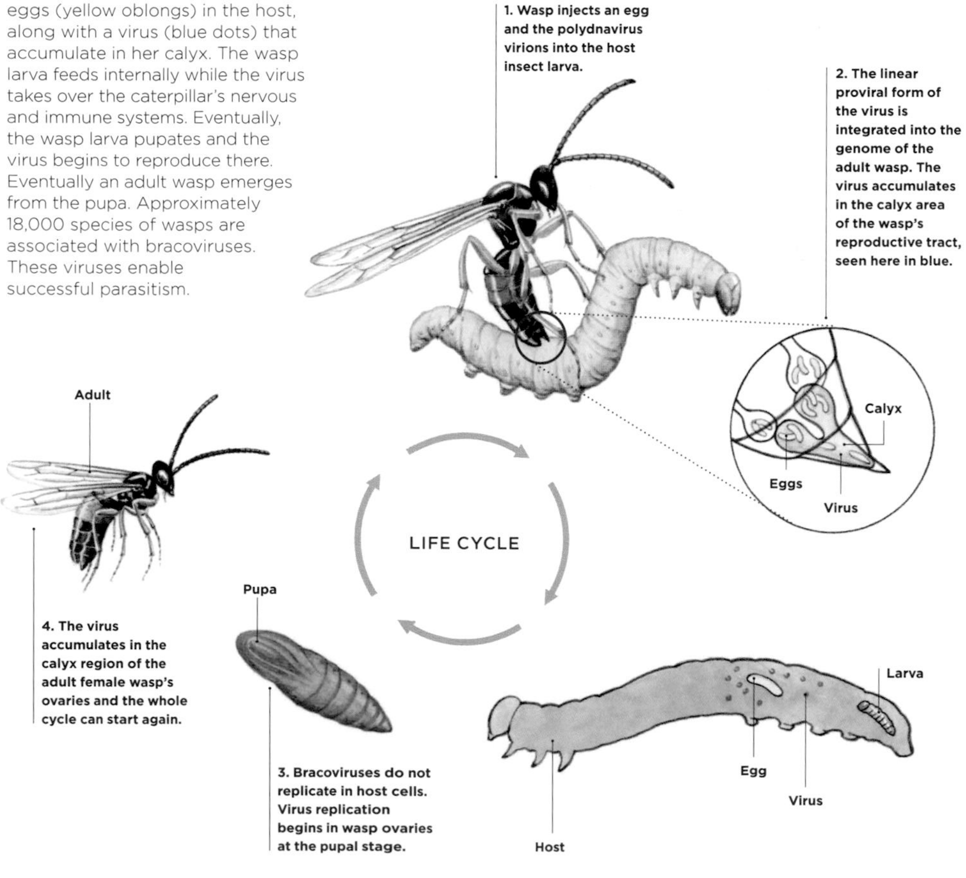

1. Wasp injects an egg and the polydnavirus virions into the host insect larva.
2. The linear proviral form of the virus is integrated into the genome of the adult wasp. The virus accumulates in the calyx area of the wasp's reproductive tract, seen here in blue.
Calyx
Eggs
Virus
Adult
LIFE CYCLE
Pupa
4. The virus accumulates in the calyx region of the adult female wasp's ovaries and the whole cycle can start again.
Larva
Egg
Virus
Host
3. Bracoviruses do not replicate in host cells. Virus replication begins in wasp ovaries at the pupal stage.

Agricultural Ecosystems

Croplands present artificial ecosystems markedly different from natural ones, as monocultures favor a few species while excluding most others. Some privileged species become pests, but others, termed "beneficials," are allies. Accidental and intentional introductions of foreign species have complicated matters.

One severe agricultural pest is the wheat stem sawfly, *Cephus cinctus*. In North America, this species causes annual losses in excess of 350 million U.S. dollars. Since the larva bores into the hollow stems, chemical treatments are ineffective. Two braconid wasp species in the genus *Bracon* are natural enemies, but not ideal agents of control. Destroying crop stubble in late autumn or early spring helps, as does switching to solid-stem varieties of wheat, although the yield potential is lower.

There exist a surprising number of "seed chalcids," wasps in the family Eurytomidae (and a few in Torymidae) that attack crops all over the world, especially nut and fruit trees. *Eurytoma amaygdali*, the almond fruit wasp, is a major pest of orchards in the southeast Mediterranean region and Middle East. Torymids in the genus *Megastigmus* infest seeds of orchard trees and conifers. The pistachio seed chalcid attacks that crop in northern Africa, the Mediterranean, and Middle East. The Douglas-fir seed chalcid, *Megastigmus spermotrophus*, and its relatives, are enemies of conifer seed orchards. In contrast, the Brazilian peppertree seed chalcid, *Megastigmus transvaalensis*, is a potential biocontrol of its host in Florida, U.S.A, where this South American tree is invasive.

The first natural enemy imported to control an invasive pest was *Cotesia glomerata* (family Braconidae), brought to the U.S.A. from Europe in the 1880s to combat imported cabbageworm, *Pieris rapae*. The first commercial application of a parasitoid was probably *Encarsia formosa* (family Aphelinidae) against whiteflies on greenhouse tomatoes in the U.K. in the 1920s. Advances in insecticide efficacy in the 1940s pushed research on parasitoids into the shadows until the 1970s, when the public rebelled against DDT. The most astounding success story of parasitoid control has taken place in Africa. Cassava is a dietary staple there, but in 1973, a mealybug outbreak began decimating the plants. Researchers turned to a wasp from Central America where cassava is also native, but trials failed. It turned out that two different mealybugs were on the plants, and the worst pest was from South America. Enter *Apoanagyrus lopezi* (family Encyrtidae) from Paraguay. It yielded stupendous results in Nigeria, and by 1990 the wasp was in use in 26 African nations covering an area of more than 2.7 million square kilometers. To this day, in over 95 percent of the region, the wasp has achieved permanent control of the mealybug, saving of at least 200 million U.S. dollars.

Nut Pest
The seed chalcid *Eurytoma amygdali*, family Eurytomidae, is a serious pest in almond orchards in the Middle East and parts of the Mediterranean.

To The Rescue
Cassava crops in Africa were plagued with mealybugs from South America until an encyrtid wasp, *Apoanagyrus lopezi*, was discovered to be a lethal enemy of the pests. The wasp has saved many Africans from famine.

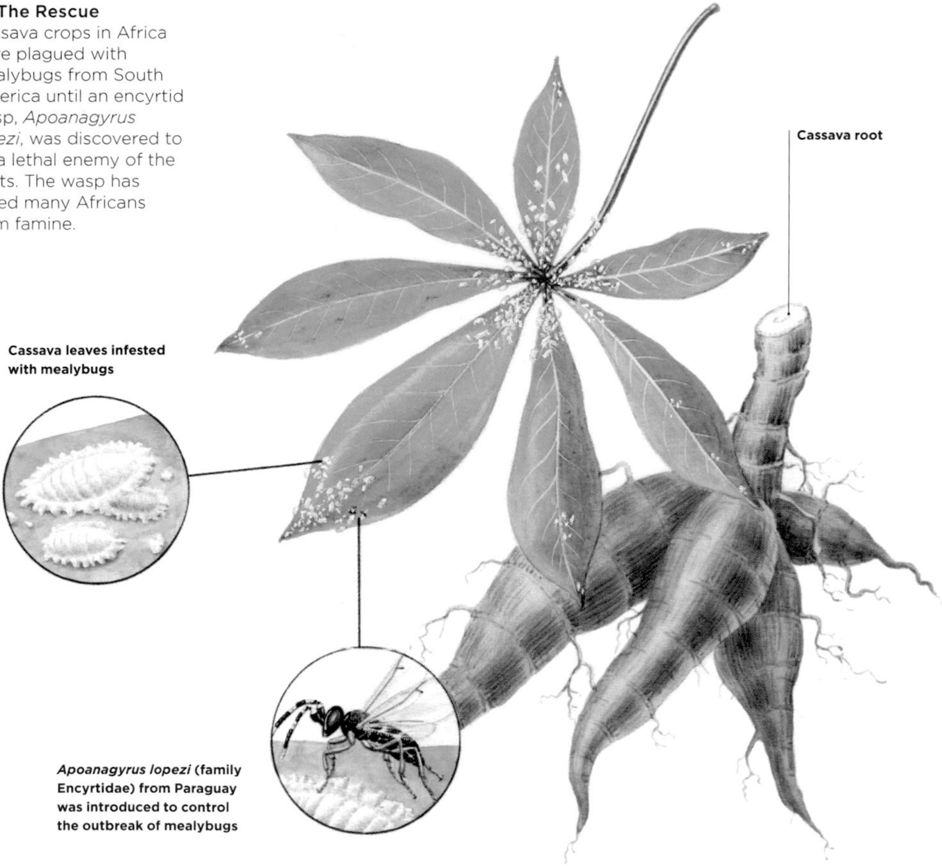
Cassava root
Cassava leaves infested with mealybugs
Apoanagyrus lopezi (family Encyrtidae) from Paraguay was introduced to control the outbreak of mealybugs

Cereal Pest
Some stem sawflies, like this *Cephus pygmaeus*, bore as larvae in the stems of wheat and other cereal crops. Because they are inside the plant, the larvae are not easily killed with pesticides.

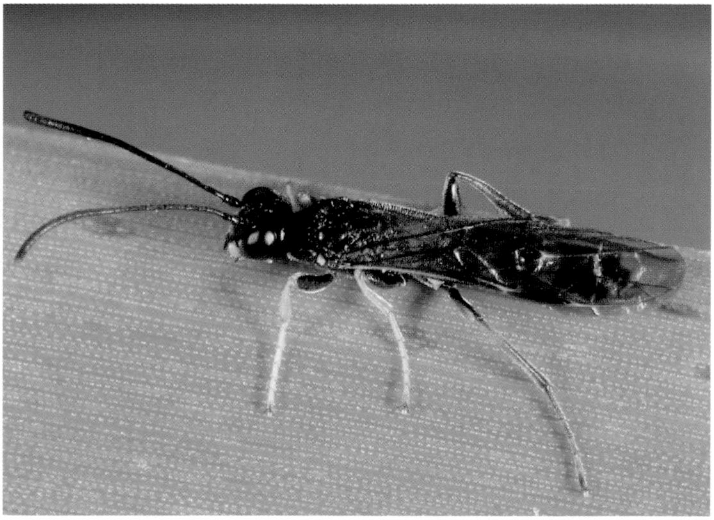

PROFILE

Miniature Warriors

Trichogramma spp.

Good things come in small packages, especially if they are nearly invisible wasps in the genus *Trichogramma*—egg parasitoids of moths, true bugs, and a few other insects. So efficient are these miniature warriors that they are mass-produced for use in fighting agricultural pests, and even sold in retail outlets for your home vegetable garden.

Egg Parasitoid
Tiny wasps in the genus *Trichogramma* are reared in great quantities on eggs of the angoumois grain moth. The wasps are then released to combat many other agricultural, stored grain, and garden pests.

The genus *Trichogramma* was formally described in 1833 by the English entomologist John O. Westwood. There are currently over 200 known species, most of which have been named since 1960. In all species, each individual wasp undergoes metamorphosis within a single egg of the host insect. One female wasp can parasitize up to ten host eggs per day.

The female wasp carefully locates the appropriate host egg by interpreting mostly olfactory and visual stimuli. She drums her antennae on the surface to determine whether the egg is already hosting a parasite, and to assess how many of her own offspring the egg could sustain. The duration of the life cycle depends on the host, and air temperature. The hotter it is, the shorter the time needed to complete metamorphosis. It can take as little as a week on average, but can be as short as 144 hours at over 90° Fahrenheit, as one study showed.

Family Trichogrammatidae

SPECIES	>839
DISTRIBUTION	Worldwide, except for Antarctica
SIZE	0.1–0.47 inches (0.3–1.2mm)
AMAZING FACT	Some tiny species have fewer than 10,000 neurons

Wasp Hosts
The angoumois grain moth is an example of a host used to rear *Trichogramma* in large numbers at insectaries. The adaptability to different hosts is both a blessing and curse in biocontrol, as other beneficial insects may be parasitized.

It was in the early 1900s that *Trichogramma* were first bred in captivity for use as biocontrol agents. The first mass-rearing enterprise involved *Trichogramma minutum*, the common European species, in 1926. The host used was the angoumois grain moth, *Sitotroga cerealella*, itself a pest of stored grain. The adaptability of "trichos" to alternate hosts makes them excellent candidates for captive breeding. While there was interest in *Trichogramma* in the U.S.A., championed chiefly by entomologist Stanley E. Flanders in a 1930 paper in *Hilgardia*, post-Second World War emphasis on new chemical pesticides eclipsed the idea of biocontrol. China started developing mass-rearing in 1949, and by the 1960s Europe and the U.S.A. rekindled their research. The 1970s saw mass-rearing and releases into sugarcane and corn. Between 1975 and 1985, pests of cotton, sugarbeets, vineyards, cabbage, plum, apple, tomato, and rice were also targeted. The saturation of fields (an "inundation release") is now common practice in 30–50 countries, with nine species of *Trichogramma* in regular use.

Angoumois grain moth, *Sitotroga cerealella*.

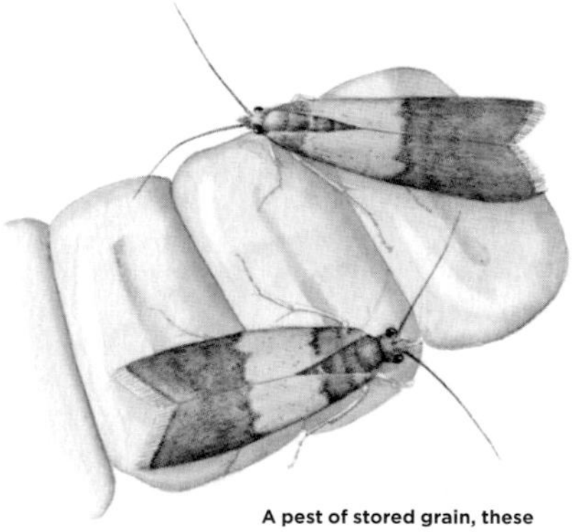

A pest of stored grain, these moths lay their eggs in the grain, thus rendering it agriculturally unusable.

Too Tiny
An "actual size" illustration is impossible, so small are these wasps. The period at the end of this sentence is a better measure of their enormity.

Actual size

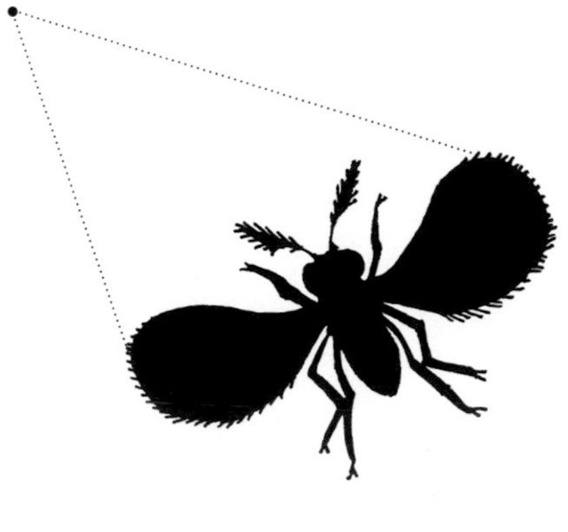

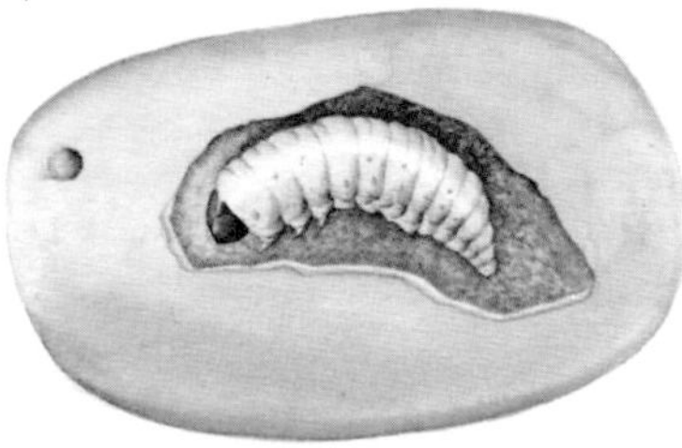

Trichogramma were mass-reared as parasitoid pest control, thus ending propagation and the presence of larvae in the grain.

Urban and Suburban Ecosystems

Even the inner city is not devoid of wasps, but they are abundant and diverse in residential neighborhoods. Most evident may not be the insects themselves, but what they create. We may be shocked to discover a sizeable, abandoned nest of aerial yellowjackets in the shrub by the porch, exposed when the foliage falls in autumn—they were there all along but we failed to notice.

Potter Wasp
In India, a female *Phimenes flavopictus* flies a ball of mud to her nest. She will use it to seal her pot, already filled with paralyzed caterpillars for her single offspring to eat.

Our homes provide a unique and attractive residential environment for wasps. Mud daubers and potter wasps plaster their nests in door jambs, recessed window frames, and up in the attic if we have not excluded them from entering the vents. Paper wasps suspend their nests from under eaves, or inside hollow pipes we forgot to cap. The birdhouse we neglected to maintain and the overturned flower pot are also fair game.

Mason wasps plug the wind chimes, finding the pipes to be perfect cavities for nesting, if there are no beetle borings in the unfinished fence rails. In North America, grass-carrier wasps of the genus *Isodontia* have taken to exploiting the tracks of sliding windows for their nesting needs. Homeowners are puzzled to find grass, "worms," and paralyzed tree crickets in the frame. In the garden, we discover digger wasps tunneling away, sometimes in the wasp equivalent of a subdivision. A settlement of dozens of huge cicada killers (*Sphecius* spp.) can contribute to your procrastination of yardwork.

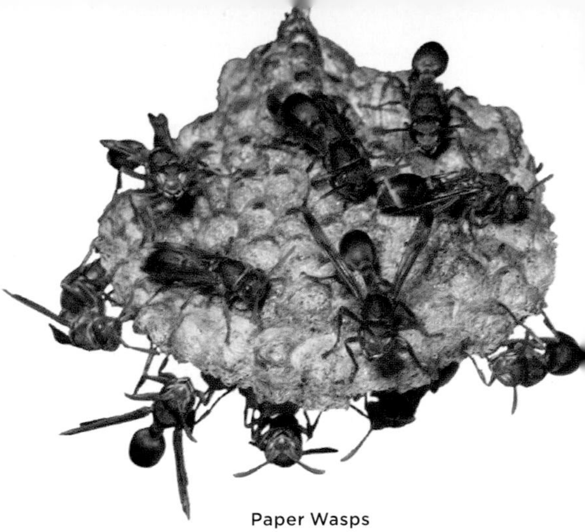

Paper Wasps
Paper wasps, like this colony of *Ropalidia marginata*, often build their nest on houses, buildings, and other structures in urban and suburban areas. Most go unnoticed by people.

Sawfly larvae are eating your roses, and the mossy rose gall wasp, *Diplolepis rosae*, makes the bush look even worse. The storm-damaged hardwood tree is now attracting big, droning horntail wasps, followed by their equally enormous ichneumon wasp parasitoids. The ornamental pine has a sawfly infestation, too, and aphids and scale insects are spurting honeydew that draws even more wasps. Fortunately, your local wasps are hunting caterpillars, grasshoppers, turf grubs, or other pests that would ravage your garden, or killing flies that might spread bacteria at your backyard barbecue. At night, nocturnal ichneumon wasps, braconid wasps, and others are drawn to the porch light, distracted from hunting their hosts by the bright beacon.

Inside your home, there are still more wasps. Solitary female ensign wasps creep about looking for cockroach egg capsules in which to lay their own eggs. The moth caterpillars and beetle larvae infesting the dry foods in your pantry are hunted by tiny bethylid wasps that sting the pests into momentary paralysis, lay a single egg on each, and leave. The bethylid larva that hatches then feeds as an external parasitoid, eventually killing its host.

Wasps are everywhere, but there is no need to panic. They are here to help.

Mud Daubers
One female mud dauber can construct many cells in her lifetime. Here, nests of *Sceliphron curvatum* fill crevices in a brick wall in Italy. Old cells are spilling fragments of pupa cocoons (brown).

PROFILE

Ensign Wasps

Family Evaniidae

Ensign wasps are named for the tiny, flag-like abdomen they wave up and down as they walk. They make up the family Evaniidae and are found mostly in the tropics. There are roughly 14 genera and 400 species known. You may recognize the blue-eyed *Evania appendegaster* even if you rarely venture outdoors. That is because this little wasp is a predator of cockroach egg cases, like all evaniids. You might also mistake this wasp for an ant, as it rarely flies.

Blue-eyed Heroine
The turquoise eyes of *Evania appendigaster* help to identify these ant-like wasps should you encounter one. They frequent the *indoors*, hunting for the hidden egg pods of cockroaches.

The hosts for this wasp are among our most notorious household pests: The American cockroach, *Periplaneta americana*; the Australian cockroach, *Periplaneta australasiae*; and the Oriental cockroach, *Blatta orientalis*. All three, though common in cities around the world, are believed to be native to Africa and/or the Middle East. The common names assigned to these species tend to reflect mostly derogatory political sentiments of the past. Strangely, the origins of the ensign wasp are a mystery, but are suspected to be Asian.

Upon discovery of a viable host egg case (ootheca), the female *Evania appendigaster* takes 15–30 minutes to penetrate the tough shell with her ovipositor. She will insert only a single egg, which hatches in one to three days. The larva that hatches is a predator that acts like an internal parasitoid, using its strong, heavily-armored jaws to crunch through each roach egg within the capsule. Should more than one wasp deposit her egg in the same ootheca, cannibalism is the usual outcome. According to recent laboratory studies, the wasp larva progresses through three instars in 14–36 days, before resting as a prepupa for an additional 17–26 days. Most feeding and growing occurs in the first two instars. In temperate climates, overwintering likely occurs as a prepupa. The pupa stage lasts between 21 and 40 days, after which the adult wasp emerges and chews its way out of the ootheca to freedom.

Family Evaniidae

SPECIES	~449
DISTRIBUTION	Worldwide except for Antarctica
SIZE	0.1–0.27 inches (3–7mm)
AMAZING FACT	Larvae are predators of cockroach eggs

Actual size

Half a Wasp?
The abdomen of ensign wasps is so small that it often appears to be missing. The wings fold roof-like over the back, making the wasp compact enough to enter narrow crevices.

Despite the attributes and efficiency of this cockroach egg predator, there has been no concerted effort to employ it as a pest control agent in domiciles. More research is needed to reconcile historical differences in the account of the wasp's life history, and to identify the most appropriate densities of the parasitoid for maximum effect.

Egg Seeker
A female *Evania appendigaster* clambers over a dead American Cockroach while looking for a roach egg pod (ootheca) to lay an egg in. Her larva will eat all the eggs in that pod.

6

Diversity
Success Through Variation

Exploiting Every Opportunity

Why are wasps so diverse and successful? Because they are masters at partitioning resources and they are capable of complex behaviors, including social nesting. Even excluding ants and bees (which are themselves wasps), wasps in the traditional sense are close to the pinnacle of advanced insect evolution.

Take a hypothetical plant as an example of a resource used by wasps. Sawfly larvae, like caterpillars, chew leaves, while other wasps are leaf miners, some bore in the stem and gall wasps form galls. Many wasps feed on flower nectar, and a few harvest pollen to feed their young. One plant thus sustains a great variety of wasps.

This scenario also applies if you replace the plant with, say, a moth. Every stage of the moth's life cycle can be host to a wasp, often many from several different wasp families. The egg of the moth can host ultra-small egg parasitoids that complete their entire life cycle inside the moth egg. Another wasp may lay its egg in the moth egg, but the wasp larva that hatches becomes a parasitoid of the moth caterpillar, or even the moth pupa. Still other wasps deposit their eggs directly into the caterpillar, their offspring emerging from the caterpillar as adult wasps. Larvae of hyperparasitoids attack the parasitoid larvae in the ultimate game of one-upmanship.

Stinging parasitoid wasps inject venom into the caterpillar, then cart it off to a nest where it is stored as food for the wasp's offspring. A kleptoparasitic wasp invades the nest of one of the stinging parasitoids and lays her eggs there. Her larvae will then consume the caterpillar larder that was intended for the rightful occupant of the nest. Stinging social wasps simply grab the caterpillar and chew it up to take back to the nest for distribution to the larvae. Again, one species can be host to an astonishing array of wasp parasitoids and predators, both directly and indirectly.

What we know of wasp associations with host plants and insects is the result of painstaking scientific inquiry in the field and in the lab. These studies usually reflect economic concerns in agriculture and forestry. That we know as much as we do about non-pest species is a testament to the personal curiosity of many entomologists. Basic research is key, as it informs routes of specialized research that can have implications for such diverse fields as neuroscience, biosecurity, bioprospecting, medicine, biomechanics, and biomimicry. Today, citizen science projects allow anyone with patience and a small material investment to contribute to our overall knowledge of other species. Your observations can easily lead to new discoveries, right in your own backyard.

Modus Operandi
A single generic moth species can be attacked by wasps at various stages in its life cycle, especially from egg to pupa. The ability to partition resources is one aspect of wasp biology that makes them so successful.

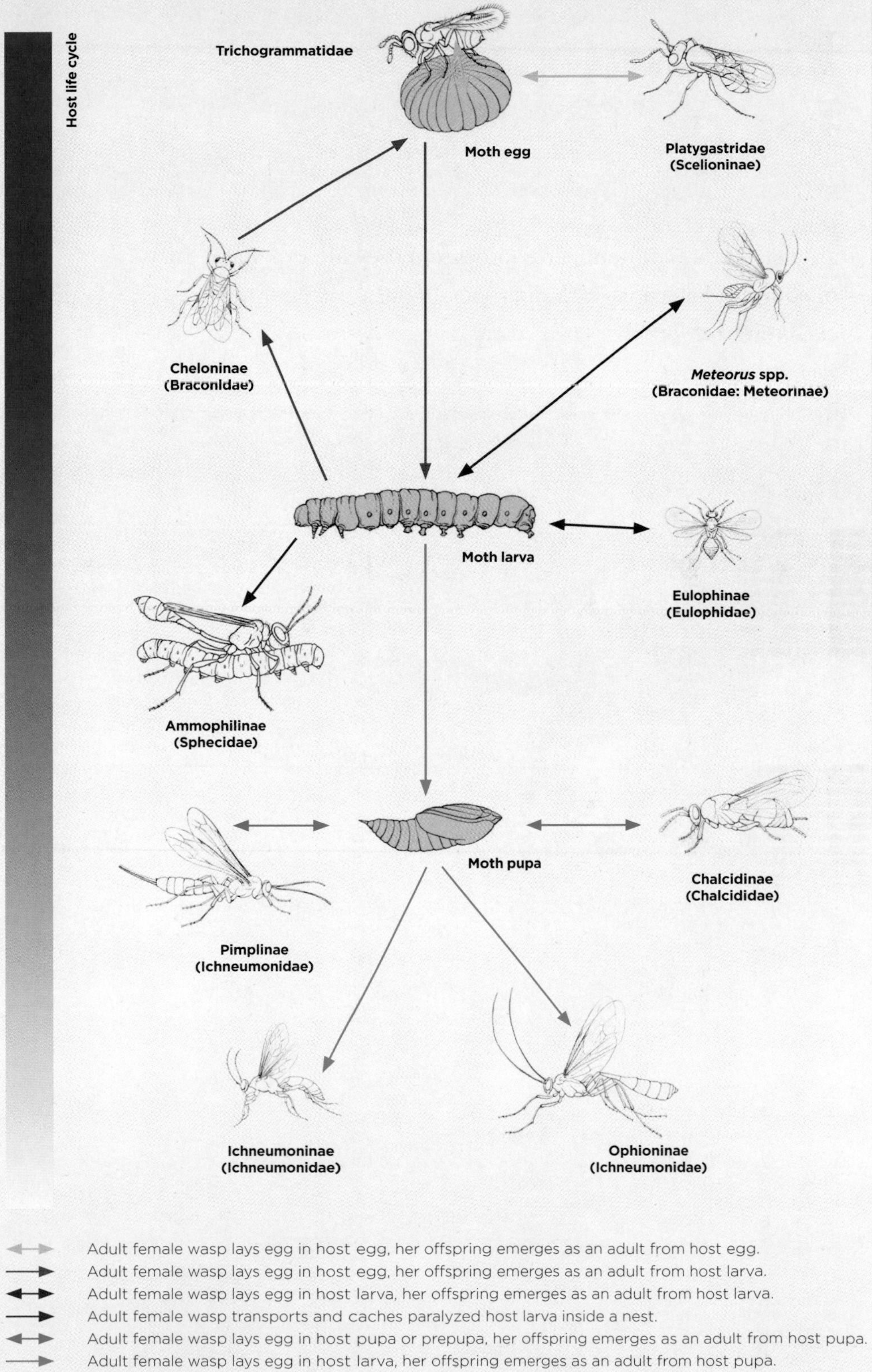
Host life cycle
Trichogrammatidae
Moth egg
Platygastridae
(Scelioninae)
Cheloninae
(Braconidae)
Meteorus spp.
(Braconidae: Meteorinae)
Moth larva
Eulophinae
(Eulophidae)
Ammophilinae
(Sphecidae)
Moth pupa
Pimplinae
(Ichneumonidae)
Chalcidinae
(Chalcididae)
Ichneumoninae
(Ichneumonidae)
Ophioninae
(Ichneumonidae)
Adult female wasp lays egg in host egg, her offspring emerges as an adult from host egg.
Adult female wasp lays egg in host egg, her offspring emerges as an adult from host larva.
Adult female wasp lays egg in host larva, her offspring emerges as an adult from host larva.
Adult female wasp transports and caches paralyzed host larva inside a nest.
Adult female wasp lays egg in host pupa or prepupa, her offspring emerges as an adult from host pupa.
Adult female wasp lays egg in host larva, her offspring emerges as an adult from host pupa.

The Vegetarians

All wasps dedicated to a strict plant-based diet as larvae are lumped in a group called the Symphyta. There are 13 families of wasps in this unofficial category, dominated by the sawflies. Symphyta resemble other wasps in color pattern, but are recognized by their lack of a "wasp waist." The abdomen is broadly joined to the thorax, so the whole adult insect is cigar-shaped. These wasps exploit a wide variety of plants in myriad ways.

Rose Sawfly
The adult rose sawfly, *Arge ochropus*, grew up as a larva feeding on rose leaflets. Its diet sometimes makes this species a pest.

Collectively, sawfly larvae feed on everything from ferns to conifers. No one wants their favorite rose bush consumed by sawflies, but would it help to know that another species (*Arge humeralis*) eats poison ivy? Sawfly larvae resemble caterpillars, but have at least six pairs of prolegs, the fleshy "knobs" on the abdominal segments—true caterpillars have no more than five pairs. A few sawfly larvae appear slimy, like the rose slug. A number of species sequester the toxins of their host plants, or manufacture their own compounds for defense. Bother a sawfly larva and it will likely coil up, rear up, and/or spit disgusting fluids from its mouth, anus, or both. Some members of the families Pergidae and Argidae are lethal to livestock that graze or forage and ingest these larvae. Massive casualties have occurred in Australia, Denmark, and Uruguay, where animals appear to become addicted to eating the insects.

Foliage, buds, fruits, shoots, stems, and catkins are all utilized by at least one kind of sawfly. Horntails, xiphydriid woodwasps, and incense cedar wasps (Anaxyelidae), bore in wood as larvae. Most larvae of Symphyta munch leaves or conifer needles in plain view, either as single individuals or in groups, but there are plenty of exceptions. Some are leaf miners, concealed between the layers of a leaf, and leaving unsightly blotches of dead tissue in their wake. A few sawflies form galls on leaves, on leaf petioles, or stems. Some are leaf folders or leaf rollers. Webspinning sawflies in the family Pamphiliidae include leaf rollers, but some spin communal webs under which they feed, much like tent caterpillars and webworms.

Apart from Symphyta, there are fig wasps (Agaonidae), "seed chalcids" in the families Eurytomidae, Torymidae, and Pteromalidae, plus seed gall-makers in Eulophidae and

Adult Butternut Woollyworm
The adult butternut woollyworm is a small, handsome insect, but not as striking as its larval stage (see above right).

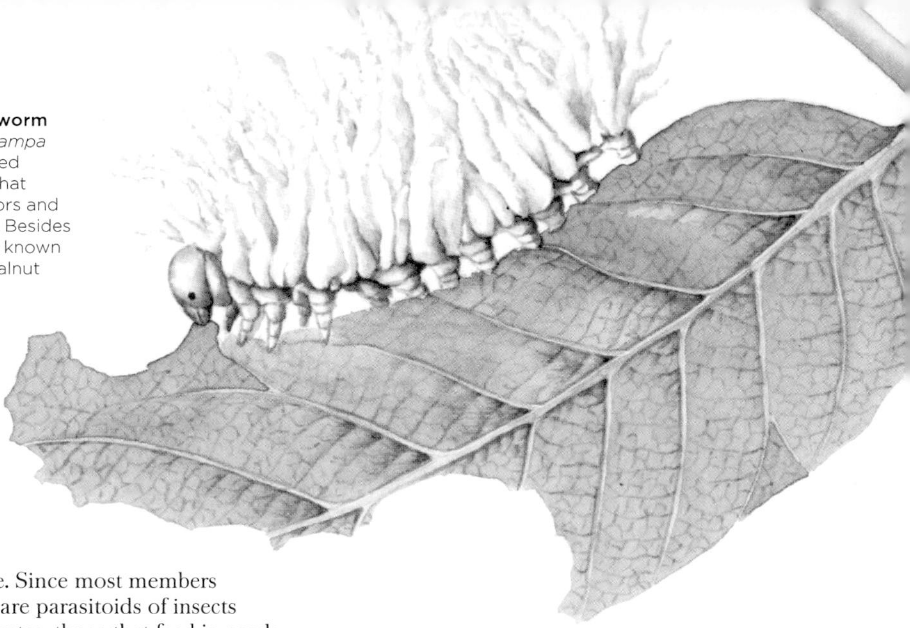

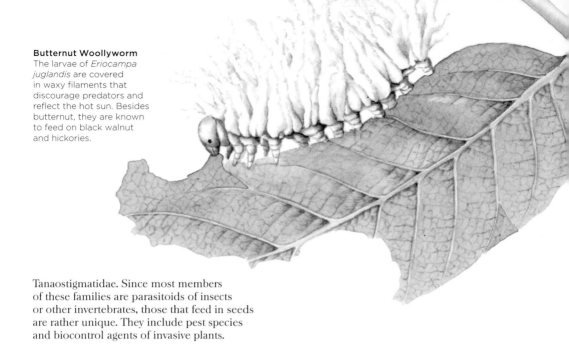

Butternut Woollyworm
The larvae of *Eriocampa juglandis* are covered in waxy filaments that discourage predators and reflect the hot sun. Besides butternut, they are known to feed on black walnut and hickories.

Tanaostigmatidae. Since most members of these families are parasitoids of insects or other invertebrates, those that feed in seeds are rather unique. They include pest species and biocontrol agents of invasive plants.

There is one outlier in the Symphyta: the Orussidae, which are parasitoids of wood-boring beetle larvae, horntails, and woodwasps. Orussids are common, but are seen infrequently, as they resemble ants running on logs and dead trees.

Cursive Catastrophe
The larva of an elm zig-zag sawfly, *Aproceros leucopoda* (below), feeds between the layers of a leaf. This member of the family Argidae is a pest in Asia and is spreading to Europe.

SPOTLIGHT ON

Wasps and Ants

Ants are ferocious predators of other insects, and aggressive in defending themselves and their colonies. Consequently, most animals avoid them, but a number of wasps take advantage of ants to achieve protection, exploit helpless ant brood, or even prey on adult ants.

The tiny, delicate wasps in the family Diapriidae would seem unlikely insects to successfully infiltrate ant nests, but members of at least 34 genera do so. Acanthopria and Mimopriella attack Cyphomyrmex fungus-growing ants in Central and South America. Each wasp larva is an internal parasitoid of a single ant larva. How the female wasps overcome attacks to gain entry into a nest remains a mystery. Diapriids associated with New World army ants are striking mimics of the ants, even casting off their wings, or allowing the ants to cut them off. They parade in the columns of ants, and some are parasitoids of parasitoid flies that plague the nomadic ants.

All members of the family Eucharitidae are, as larvae, external parasitoids of ant pupae. Female eucharitids avoid confronting ants, instead inserting eggs into vegetation. The larvae that hatch are active planidia that feverishly seek any moving object to glom onto. Should that be an ant, then the planidium is carried into the nest where it detaches from the worker ant and latches onto a larva. There it remains, inert, until the larval ant pupates. Only then does the eucharitid larva commence feeding.

Four species of "ant queen kidnappers" in the genus Aphilanthops, (all found in the U.S.A.), attack and sting alate (winged) queen ants of the genus Formica as they land after their nuptial flight. Up to four ants are stored per cell in an underground burrow as food for the wasp's offspring. Eight species of Clypeadon, and one species of Listropygia, confined to the western U.S.A. and adjacent Mexico, hunt worker harvester ants (*Pogonomyrmex* spp.) exclusively. The female wasp transports her comatose victim in an "ant clamp" at the tip of her abdomen. She likewise caches her victims in underground chambers.

Many New World tropical social wasps gain added protection by building their nests adjacent to the nests of dolichoderine ants. One entomologist found eight different wasp species with nests in the same tree as an Azteca ant nest.

Stealthy Strategy
A female eucharitid wasp, *Orasema* sp., lays eggs in a leaf in Colombia. The larvae that hatch stick onto passing ants, gaining entry into the nest. There, they feed on ant eggs and larvae.

Strange Invader
The wingless female of a diapriid wasp, *Bruesopria* sp., has made herself at home in a nest of thief ants, *Solenopsis molesta*, on a prairie in Kansas, U.S.A. The exact nature of the relationship remains a mystery.

Kidnapper
A female ant queen kidnapper, *Aphilanthops frigidus*, has captured and paralyzed a flying queen ant. The victim will be food for the wasp's offspring in a cell at the end of an underground burrow.

The Gall-makers

While most of the wasps in the superfamily Cynipoidea are plant-feeders, this group represents something of a bridge between vegetarian and parasitoid wasps. Galls are themselves ecosystems around which many species revolve, and it is often difficult to determine which wasp, if any, is responsible for the gall, and which are parasitoids or inquilines ("guests").

Galls are not tumors, nor in any way representative of poor plant health. One intriguing hypothesis suggests that fleshy fruits evolved from wasp-induced galls on the reproductive structures of plants. Gall-encased seeds, the theory goes, were more attractive to birds and mammals that would better disperse the seeds. Plant galls are not always of insect origin. Mites, fungi, bacteria, viruses, and nematode worms can also be responsible. The mechanisms of gall-initiation are complex, but for insects they are usually related to the egg-laying activity of the female, or the feeding of her larval offspring. The plant responds by channeling more nutrients to the site, and often enhances cell proliferation in the vicinity, creating a swelling that serves as home and food for the larval occupant.

Chances are that any gall found on an oak or related plant in the Fagaceae (beech family), or Rosaceae (roses and kin), are made by gall wasps in the family Cynipidae. Wasps in the sister families Figitidae, Lipteridae, and Ibaliidae are parasitoids of other insects. Charipidae are hyperparasitoids. Gall wasps tend to be host specific, tied to one plant or a family of related species. The wasps typically target the leaves, buds, flowers, fruits or nuts, stems and twigs, or even roots of their host. Some species create more than one kind of gall on the same plant in a bewildering alternation of generations.

Genesis of a Gall
A female gall wasp inserts her egg in the stem of a host plant. The plant will respond to the stimulation by creating a nutrient-rich gall around the egg.

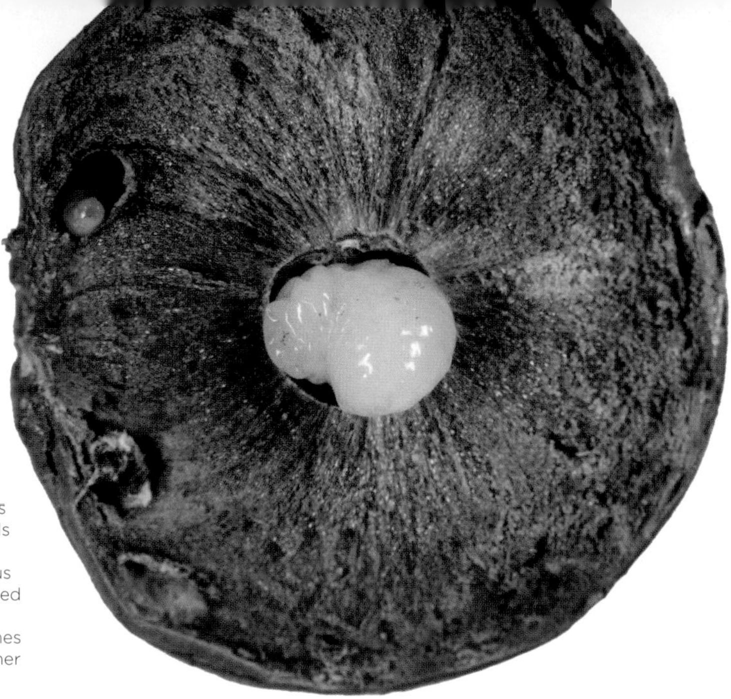

Owner, Renters
A cross-section of this oak marble gall reveals the rightful occupant larva in the center, plus larvae of an unidentified inquiline wasp on the periphery. The inquilines do no harm to the other larva in this instance.

Besides cynipids, other wasps can be responsible for galls. Some sawflies create galls on willow leaves, for example. The family Tanaostigmatidae are exclusively gall-makers, especially on seeds, so they overlap with the definition of "seed chalcids." This is also true for some genera of Eulophidae. A few members of the families Eurytomidae and Pteromalidae are also gall-formers. Most eulophids, eurytomids, and pteromalids are parasitoids, however, and if they emerge from a given gall it does not mean they created the growth.

Some cynipids make their living by ovipositing in the galls of other cynipids. Such trespassers are termed inquilines or "guests" that cause no direct harm to the rightful gall occupant, but benefit from the food and housing. An innumerable number of other wasps and insects may occupy a gall as inquilines or parasitoids. There is literally a catalog of known gall associates for the palearctic realm alone, and it is likely incomplete.

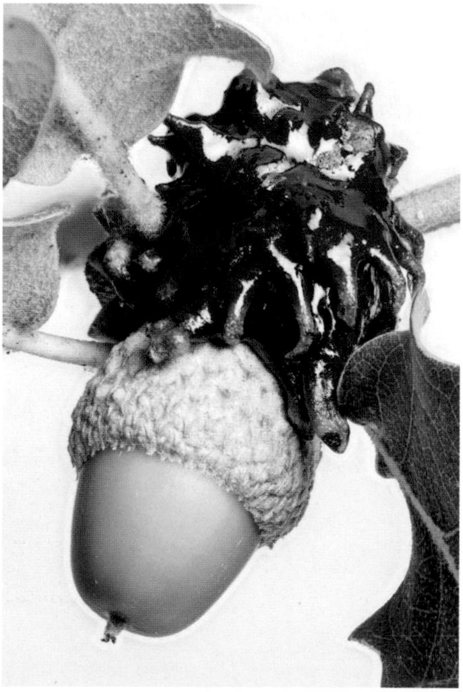

Knopper Gall
Named after the German word for a peculiar helmet, knopper galls are produced by the wasp *Andricus quercuscalicis*. The wasp alternates between *Quercus robur*, where these growths form, and Turkey oak, *Quercus cerris*, which has a different gall.

The Micro-parasitoids

Ironically, the greatest degree of diversity in Hymenoptera occurs in the smallest of wasps. Considering most species remain to be discovered or named, scientific estimates conclude there are between 833,000 and 1.1 million species of parasitoid Hymenoptera, the bulk of which are "micro Hymenoptera." They include valuable allies in agricultural and garden pest control and—bonus—they do not sting...

At least they don't sting people, anyway. Venom is more than pain-causing, paralyzing, or tissue-destroying—it includes any substance that alters the biochemistry of the host, including suppression of the immune system, and/or arrest of growth and development, that benefits the venom-inflicting organism. By that measure, most micro-parasitoids are venomous.

Micro Hymenoptera include superfamilies Proctotrupoidea, Platygastroidea, Ceraphronoidea, Mymarommatoidea, and Chalcidoidea, of which chalcidoids are by far the most diverse. The trichogrammatid wasp *Megaphragma mymaripenne* measures only 200 micrometers, comparable to single-celled organisms like *Paramecium* and amoebas. Males of the mymarid fairyfly *Dicopomorpha echmepterygis*, blind and wingless, register at 139 micrometres. A human hair is about 180 micrometres wide. In contrast, the female American Pelecinid, *Pelecinus polyturator*, a proctotrupoid, can exceed 2.4 inches (62mm).

Remarkably, some of the tiniest wasps are egg parasitoids of aquatic insects. A number of egg parasitoids, both aquatic and terrestrial, are phoretic: the female wasps ride around on the female host until she lays her eggs, at which time they disembark to parasitize the host eggs.

The lifestyles of micro-Hymenoptera run the gamut from generalists to specialists, from egg parasitoids to some hyperparasitoids (page 124), and internal or external parasitoids, mostly of larval insects. Some Platygastridae and Encyrtidae females lay a single egg that results in dozens, if not hundreds or thousands of embryos in a phenomenon known as polyembryony. "Seed chalcids" are also parasitioids, and some chemically manipulate their plant host into redirecting more nutritients to unfertilized ova for the benefit of the wasp larva inside each seed. Eurytomids collectively illustrate perhaps the broadest spectrum of parasitoid habits. *Eurytoma brunniventris* of western Europe is one example. The larva stage is usually a parasitoid of a cynipid gall wasp, but can be a parasitoid of the inquiline gall wasp *Synergus*; be a hyperparasitoid of other chalcidoid wasps; or even feed on the gall tissue.

Name any insect or insect relative on land or in fresh water, and it likely has at least one micro-parasitoid enemy. Caterpillars of moths and butterflies, and larvae of sawflies and beetles are heavily exploited, but so are scale insects, aphids, some flies, other wasps, even spider eggs. The curious naturalist could forever be sucked down the rabbit hole of research on these astonishing insects.

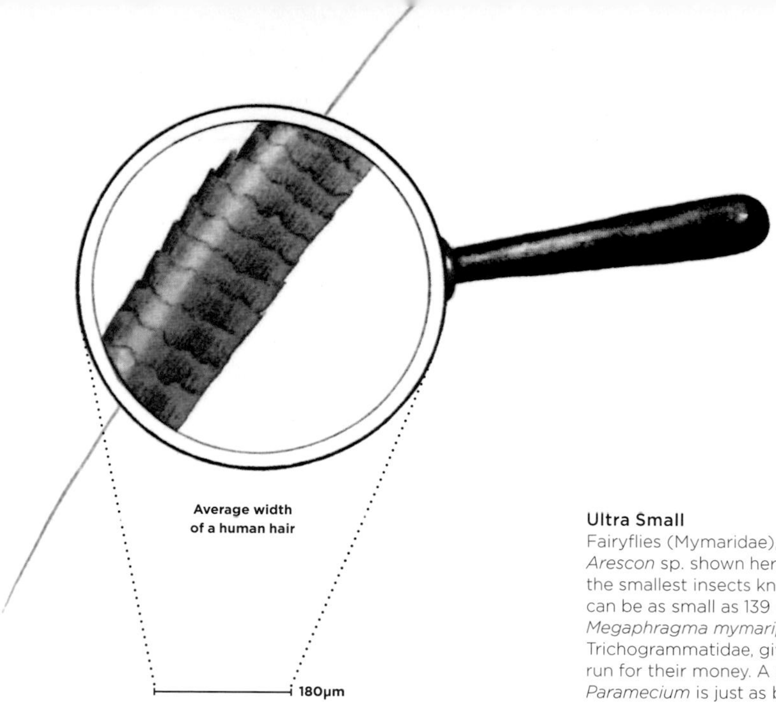

Ultra Small

Fairyflies (Mymaridae), like the *Arescon* sp. shown here, may be the smallest insects known. They can be as small as 139 micrometers. *Megaphragma mymaripenne*, family Trichogrammatidae, gives them a run for their money. A single-celled *Paramecium* is just as big.

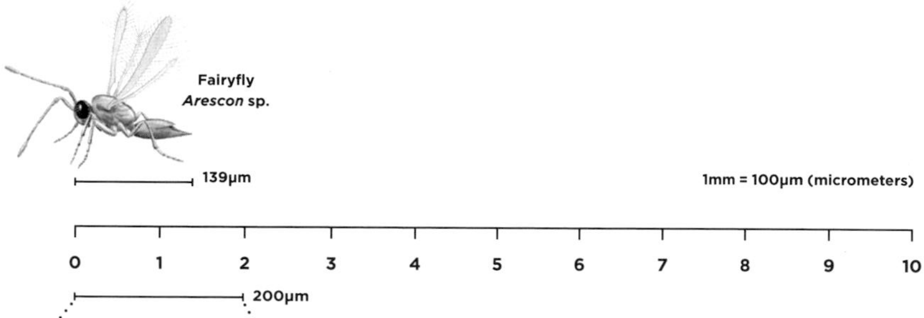

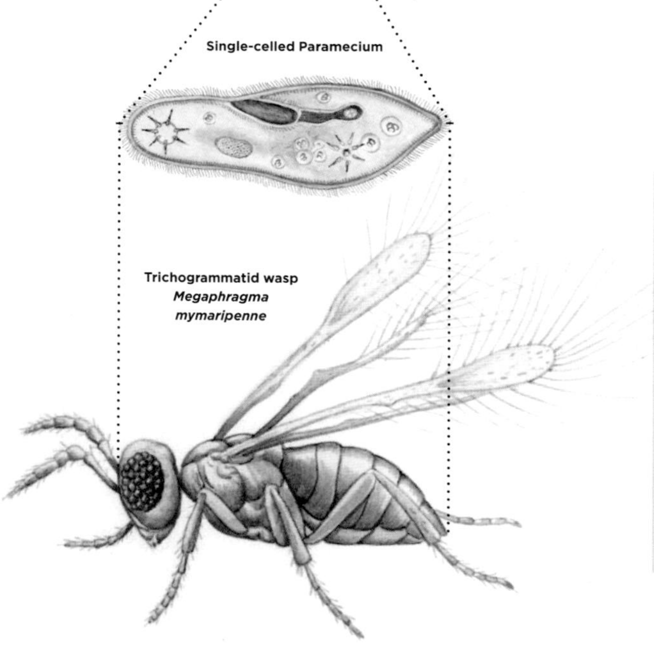

Comparative Sizes

100µm

= 0.1mm

= 0.00394 inches

= smallest distance that can be seen with the naked eye

= thickness of a coat of paint

= length of a dust particle

500µm

= average length of a grain of sand

= average length of a grain of salt

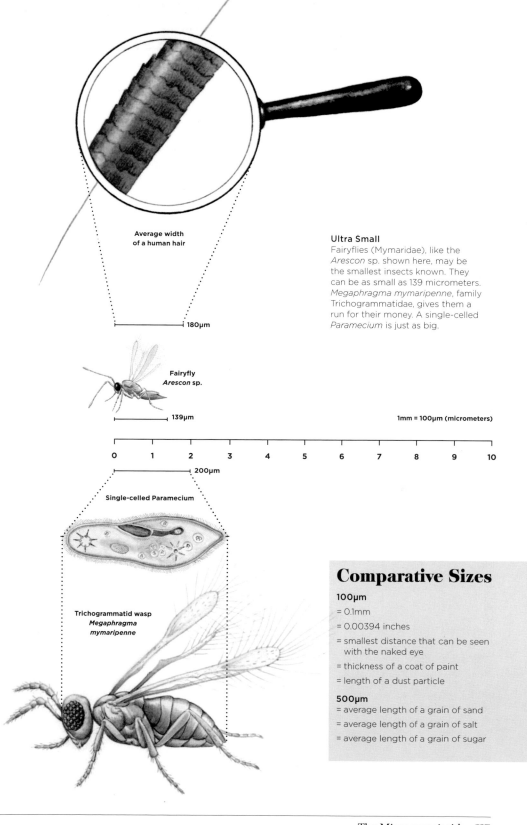

= average length of a grain of sugar

The Macro-parasitoids

Micro- and macro-parasitoids belong to the suborder Apocrita, which includes all wasps except the suborder Symphyta. Their lifestyle strategies are similar, but larger wasps attack larger hosts. The most conspicuous "non-stinging" parasitoids belong to the superfamily Ichneumonoidea (families Braconidae and Ichneumonidae).

Aphid "Mummies"
Aphids victimized by braconid wasps in the subfamily Aphidiinae become bloated, moribund "mummies." Each aphid hosted a single wasp larva. The adult wasps have since emerged through the round "lids."

Stinging behavior is relative, and even micro Hymenoptera may introduce venom into the host. This is true of many Ichneumonoidea, and some larger species can sting people in self-defense. This should not warrant our hostility, as these insects are vital in keeping true pests at bay.

Next time you find aphids, look for ones that appear bloated and gray or beige. Those are "aphid mummies," each consumed from within by a tiny braconid wasp larva (subfamily Aphidiinae). The wasp pupa is inside, unless there is a big hole in the mummy, in which case the newly-minted adult has departed. Braconids in the genus *Aleiodes* mummify caterpillars, thickening the "skin" of their deceased host to protect for the wasp's pupa stage. Multiple wasp larvae exit, leaving "stigmata" holes behind.

Parasitoids are persistent and athletic. When attacked by the braconid wasp *Diolcogaster facetosa*, a green cloverworm, (*Plathypena scabra*), launches itself from the plant, releasing a fine silk strand on which it safely descends like a bug bungy-jumper. This does not deter the wasp, which deftly shimmies down the thread to reach the caterpillar anyway.

Amazingly, the braconid *Cotesia glomerata* uses volatile compounds released by damaged plants to locate its host. Cabbage nibbled by a butterfly caterpillar sets off aromatic alarm bells that recruit the wasps to come to the plant's rescue. Caterpillars still munching, but now with *Cotesia* larvae inside them, cause the plant to release a different aroma that attracts a hyperparasitoid ichneumon wasp, *Lysibia nana*, that attacks the braconid.

Head Start
A female braconid, *Chelonus insularis*, injects her egg into the egg of an owlet moth. Her offspring gets a head start on other parasitoids, and will emerge as an adult from the moth caterpillar.

Collectively, most ichneumon wasps target the larvae and pupae of moths and butterflies, flies, beetles, or other wasps. A few specialize on other insects with complete metamorphosis, others feed on spiders. Larvae of some ichneumons are internal parasitioids, others external.

Like braconids and other parasitoids, Ichneumonidae use two strategies of attack. In some species, especially those targeting older host larvae, or pupae, the female wasp oviposits in the host and injects chemicals that halt further growth and development. This is the lifestyle of an idiobiont. Hosts of idiobionts typically include wood-boring larvae, or pupae inside of cocoons. Alternatively, females of other species allow the host to continue growing, and their larvae are koinobiont parasitoids that feed and grow slowly, accelerating when the host enters the prepupa or pupa stage. Hosts of koinobionts are generally more exposed, or half-concealed.

Gregarious
A caterpillar has played host to larvae of a braconid wasp in the genus *Cotesia*. Several larvae have fed as internal parasites, then erupted to spin cocoons. An adult *Cotesia* sits among them.

The Pollen-eaters

Bees are the dedicated pollen-feeding branch of the order Hymenoptera, but other wasps harvest pollen for their offspring, as well. The pollen wasps in the subfamily Masarinae are exclusively pollen-eaters in the larval stage. These wasps are solitary members of the family Vespidae that includes the social hornets, yellowjackets, honey wasps, and paper wasps, as well as potter and mason wasps.

Multi-tasking
A female pollen wasp, *Pseudomasaris vespoides*, gathers pollen from a *Penstemon* (beardtongue) blossom in Colorado, U.S.A., as a male strokes her with his long antennae, hoping to mate.

Pollen wasps are most diverse in southern Africa, but they occur in many arid climates, especially in the Mediterranean region, parts of Australia, the western third of North America, southern South America, and in scattered locations in Eurasia, north and west of India. There are 220 species.

Instead of collecting pollen in special brushes or pollen baskets like most bees do, masarines ingest pollen and nectar, storing it in the crop (a sort of ancillary stomach) for transport back to the nest. Many pollen wasps excavate multicellular burrows in the soil, sometimes adorned with a vertical or horizontal mud turret at the entrance. *Quartinia vagepunctata* lines its burrow and turret with silk spun from the female wasp's mouth. Other masarines make free-standing mud nests attached to stones, cliff faces, plants, or inside a pre-existing cavity. These mud-builders use water or nectar to cement their nest material.

Like bees, pollen wasps can be generalists, visiting a variety of flowers, or specialists that restrict themselves to only one or a few types of floral hosts. Among the specialists are some *Pseudomasaris* species of the American west that visit *Penstemon* flowers almost exclusively. The males patrol patches of flowers and quickly attempt to mate with females that have plunged head first into the narrow blossoms.

The only other wasp known to provision its nests with pollen is *Krombeinictus nordenae*, a solitary crabronid wasp found in Sri Lanka. This wasp nests in hollow stems of the tree-like legume *Humboldtia laurifolia*. Astonishingly, it appears the female wasp rears only one larval offspring at a time, bringing in pollen and nectar (from the same species of plant the nest is in) as needed. She takes the mature larva to the bottom of the plant internode where it spins a cocoon and pupates. The mother wasp then starts the cycle over with another egg near the opening of the nest chamber.

Brachygastra is a genus of social polistine paper wasps known as "honey wasps" for their habit of storing quantities of honey. The 17 species range collectively throughout Central and South America, north to the border with the U.S.A. The stores, which include honeydew from true bugs, as well as nectar-based honey, get the colony through periods of unfavorable foraging conditions. Pollen is a negligible part of their diet.

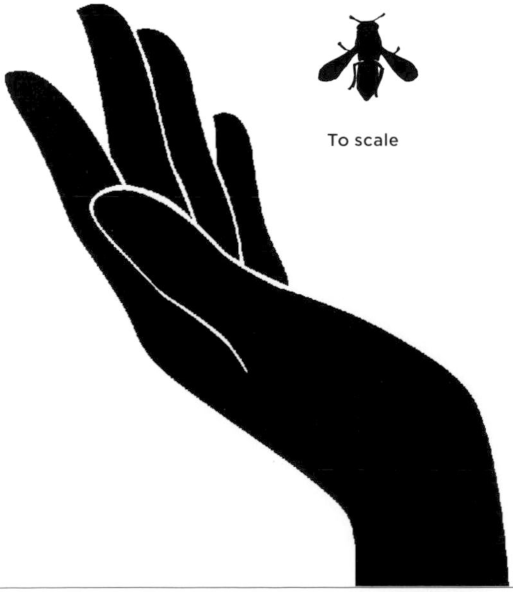

To scale

All In
A female *Pseudomasaris vespoides* dives head first into a *Penstemon* flower to reach pollen and nectar, which she ingests and stores in her crop. She'll regurgitate it into a mud cell as food for a larval offspring.

The Stinging Parasitoids

Female solitary wasps use their sting to incapacitate their hosts, and only rarely in self-defense. Together with social wasps, ants, and bees, the stinging parasitoids make up the Aculeata, a division of the suborder Apocrita. What sets these wasps apart is the repurposing of the ovipositor (egg-laying organ) into a venom-channeling sting.

Up and Away
A female European beewolf, *Philanthus triangulum*, has paralyzed a honey bee with her sting and is now transporting it to her nest burrow. There it will serve as food for one of her offspring.

Venom of solitary stinging wasps is geared to paralysis of a host insect or spider to facilitate egg-laying and, often, transport of the victim to a nest where it will be stored as food for the wasp's offspring. More "primitive" aculeates paralyze a host temporarily, attach an egg to it, and leave. The host regains mobility and resumes its life, until the wasp larva consumes it.

Hunting and nesting behavior becomes increasingly complex among the "advanced" solitary aculeates. Many unrelated families have arrived at similar destinations, such as underground burrows, pre-existing cavities, and mud constructions as nesting solutions. These wasps execute extraordinary feats of strength and agility: a female wasp can haul prey up vertical surfaces, even if the host outweighs her, and can even fly with their paralyzed victim slung beneath them. The speed at which a wasp excavates a burrow, or fabricates a mud nest, puts human contractors to shame. That is not to say a wasp is above stealing her neighbor's prey or usurping her nest.

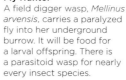

Hired Gun
The wasp *Tiphia vernalis* was brought to North America to help fight the equally exotic Japanese beetle. The wasp is a helpful biocontrol along with "milky spore" bacteria and nematode worms.

Solitary stinging parasitoids attack a variety of hosts at varying stages of maturity. Most wasps in the genus *Tachysphex* (family Crabronidae) provision their underground cells with grasshopper nymphs, while members of the genus *Prionyx* (Sphecidae) use adult grasshoppers. The bulk of stinging parasitoids are specific in their host associations, but not necessarily at species- or even family-level. The great golden digger, *Sphex ichneumoneus*, uses katydids (Tettigoniidae) as food for her offspring, but also tree crickets (Gryllidae), or raspy crickets (Gryllacrididae).

Few stinging parasitoids have been employed as agents of pest control, but the spring tiphia, *Tiphia vernalis*, was imported to the U.S.A. from Korea during the 1920s and early 1930s to help control the Japanese beetle. The female wasp, which is active in spring, tunnels into soil to find beetle grubs. She stings the host into temporary paralysis, lays an egg, and goes on to hunt another host. Her larva will feed as an external parasitoid.

More recently, the wasp *Cerceris fumipennis* has been used as a biosurveillance tool for monitoring the status of the emerald ash borer, another invasive pest. The wasps are parasitoids of a variety of other jewel beetles in the family Buprestidae, so entomologists and volunteers intercept prey-laden females arriving at the entrances to their underground burrows, and determine the species of beetle they are carrying.

Fly Hunter
A field digger wasp, *Mellinus arvensis*, carries a paralyzed fly into her underground burrow. It will be food for a larval offspring. There is a parasitoid wasp for nearly every insect species.

SPOTLIGHT ON

Cicada Killers

Cicadas are known for their loud, throbbing "songs," which are guaranteed to induce a migraine. Thankfully, there are noise enforcers in the form of cicada killer wasps that hunt these annoying bugs. All cicada killers belong to the family Crabronidae, and are collectively distributed over much of the globe. Most familiar is the genus *Sphecius*, with 21 species. Four species occur in North America, but the bulk of diversity exists between north Africa and central Asia. There are 11 species of *Liogorytes* in South America, and one species of *Exeirus* in Australia.

Few cicada killer species have been studied, but *Sphecius* and *Exeirus* behave similarly. Female cicada killers are fossorial, meaning they are diggers. They excavate burrows in the soil that terminate in one or more underground chambers, each an abode for a single larval offspring. After preparing her nest, the mother wasp searches for a cicada. Deaf to the cicada's call, but sensitive to vibrations, she gleans limbs of trees and other likely perches, eventually blundering into a cicada. The encounter may elicit a "scream" from the hapless noisemaker, before it is silenced, paralyzed by the wasp's sting. A cicada killer is capable of flying its bulky cargo back to the nest, though prolonged dragging of the victim over the ground may be necessary. One to four paralyzed cicadas are put into each underground cell, with an egg laid on the last one. The chamber is sealed with packed soil and the wasp goes about provisioning the next cell.

Mug Shot
The face of a cicada killer is dominated by the two compound eyes, yet the wasp has poor vision and detects her hosts mostly through touch as she prowls branches and twigs.

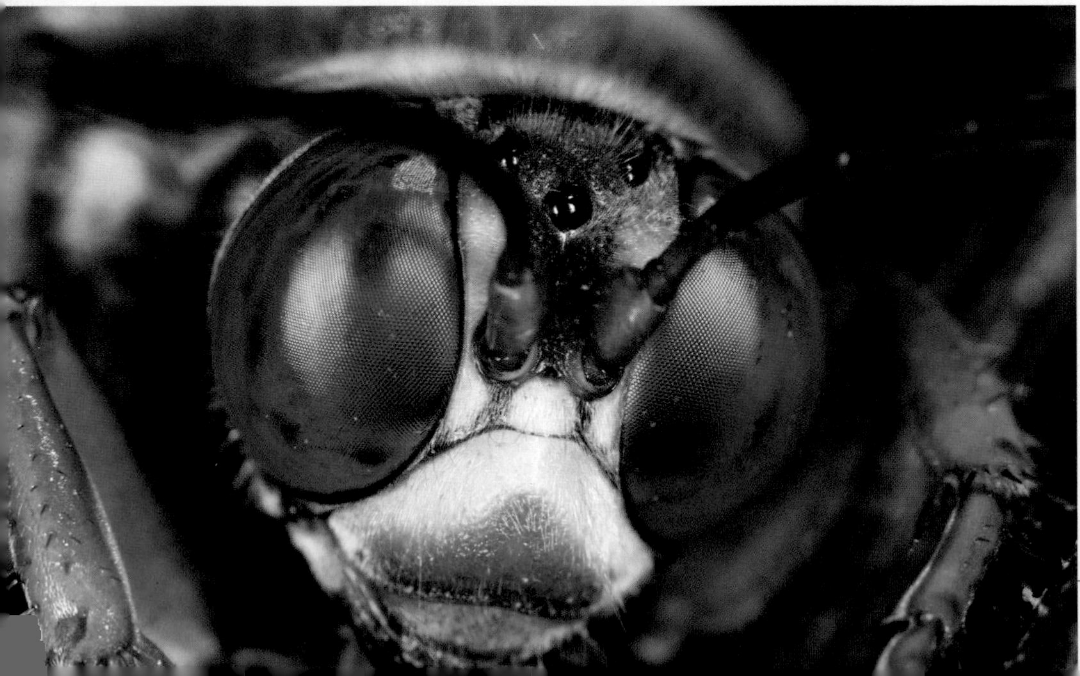

Laborious
Transporting the dead weight of a paralyzed cicada is not easy, but she can fly with her cargo, drag it up a tree, and then glide home, or drag her prize over the ground.

Cicada Crypt
The underground nest of a cicada killer contains several cells, one for each offspring. More than one cicada is needed to rear a female wasp of the next generation.

Cicada killers are solitary wasps, but it is not unusual for several female *Sphecius* to nest in close proximity in suitable soil, creating the impression of sociality. Males, though lacking a sting, are highly territorial, perching on the periphery of nest locations and chasing away all intruders.

Life is not easy for cicada killers. Birds may harass female wasps into dropping their cicadas. Satellite flies orbit the burrow entrances, waiting for a chance to deposit live maggots that will devour the cicadas ahead of the wasp larvae. Lazy cicada killers may steal cicadas from their neighbors.

Should your yard become host to nesting cicada killers, make the most of it. Hold a "watch party" to marvel at the strength and endurance of these gentle giants.

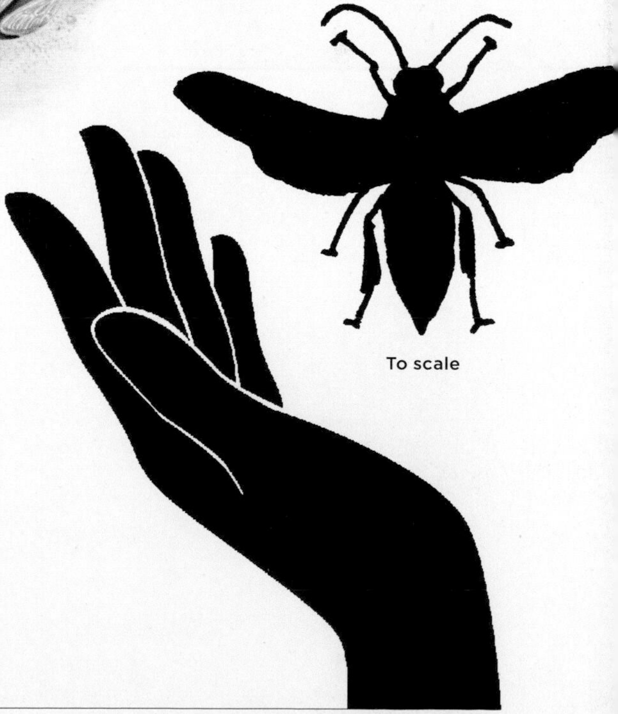

To scale

The Hyperparasitoids

Imagine the long and winding evolutionary road of wasps that are parasitoids of other parasitoids. How does that even happen? Because competition for hosts is keen, it is possible that some parasitoids evolved to attack their rivals within common primary hosts, eventually becoming dedicated hyperparasitoids. In other cases, the route to hyperparasitism is a mystery.

Hyperparasitoids exist in nearly all families of micro Hymenoptera, some Ichneumonidae, and all members of the Trigonalidae—there are 17 in total. Most hyperparasitim occurs within the immature stages of primary hosts like butterflies and moths, sawflies, or aphids, scale insects, mealybugs, and their kin. Some are obligate hyperparasitoids that can only succeed in their metamorphosis as parasitoids of other parasitoids. Others are facultative hyperparasitoids, which means they can complete their life cycle as either a parasitoid or a hyperparasitoid. There are even instances of tertiary hyperparasitism, where hyperparasitoids attack each other, including other individuals of the same species. At this wizard-level of parasitism, having options is a good thing.

Parasitoid of Parasitoid
A pteromalid wasp, possibly *Hypopteromalus tabacum*, sits atop the cocoon of a *Cotesia* sp. braconid. She will insert her egg inside, and her larva will develop on the pre-pupa of the *Cotesia*.

Perilampidae
Female perilampids lay vast numbers of eggs on foliage, and the larvae that hatch must find insects that are already parasitized by a fly or wasp larva. The perilampid larva then attacks that parasitoid.

Depending on the species, hyperparasitoids can be endoparasitoids living inside the host parasitoid, or ectoparasitoids feeding from the outside of the host. Some attack the host directly, with the female wasp ovipositing on or in the host parasitoid, while in other species the female lays her eggs in the primary host (host parasitoid's host), and her larva(e) must seek out the host parasitoid. This indirect method is prone to absolute failure if no host parasitoid is found.

Trigonalidae have a strategy with seemingly astronomical odds of success. The female wasp lays thousands of tiny eggs on foliage, destined to be ingested by a sawfly larva or moth caterpillar. Her ova hatch inside the caterpillar, but the larvae will develop no further…unless that sawfly larva or caterpillar is subsequently parasitized by an ichneumon wasp or tachinid fly, carted away by a potter or mason wasp, or killed by a social wasp. The only exceptions to this bizarre chain of events appear to be some Australian species of *Taeniogonalos* which can successfully develop in a sawfly larva without a secondary host present.

Several trigonalid eggs may be consumed by a single primary host, but only one adult trigonalid emerges from the secondary host. The genus *Bareogonalos* is restricted to hornet and yellowjacket hosts, delivered via pulverized caterpillars or sawfly larvae. In order to complete the *Bareogonalos* life cycle, the sawfly larva or caterpillar must be killed and chewed by a worker or queen wasp, then fed to the larvae inside the nest. The trigonalid larva becomes a parasitoid of the hornet or yellowjacket larva. Adult trigonalids emerge from the pupa stage of the host.

Guests, Thieves, and Killers

Ever had a houseguest who overstayed their welcome, raiding the refrigerator and otherwise making life miserable? Welcome to the world of home invaders, some more hostile than others.

Gall Community
Wasp galls support a diversity of other organisms, including other wasps. Inquilines usually do no harm, but sometimes they kill the host and become *de facto* parasitoids. Try "hatching" a gall to see what emerges.

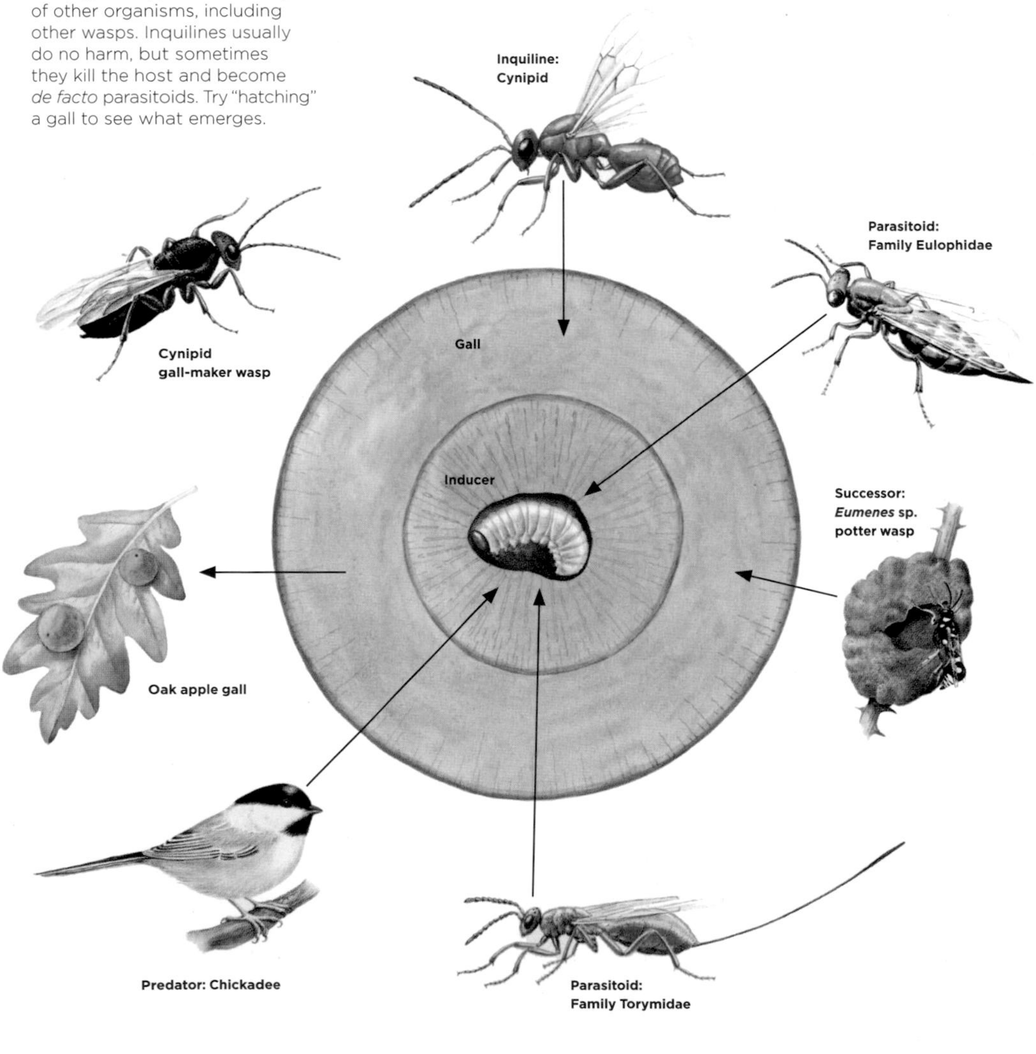

Kleptoparasitoid
The crabronid wasp *Stizoides renicinctus* of the U.S.A. digs open the nests of fossorial wasps and lays her eggs there. They will feed on the paralyzed insects intended for the rightful occupant host.

Inquilines are "guests" in the living quarters of other organisms, freeloaders that neither harm nor benefit the host. In wasps, this lifestyle is exemplified by associates of gall wasps in the family Cynipidae. The cynipid tribe Synergini includes 186 species that live this way, as do some members of the family Figitidae. Should a *Diplolepis nodulosa* rose gall be invaded by the inquiline *Periclistus pirata*, the result is a structure three times the size of the original, with up to 17 added chambers, each housing a larva of the inquiline. Gall inquilines in the strictest sense occupy the peripheral tissues of the gall. Those that share the same chamber as the larva of the gall-maker include murderers that kill the host in order to feed without competition.

Kleptoparasitoids (aka cleptoparasitoids) kill the host larva through starvation, or murder it outright like the lethal gall inquilines. These are the insect equivalents of bird "cuckoos." Many cuckoo wasps (family Chrysididae), live this way. So do sapygids (family Sapygidae), most velvet ants (Mutillidae), some spider wasps (Pompilidae: Ceropalinae and the genus *Evagetes*), and some crabronids (Crabronidae: tribe Nyssonini, and genus *Stizoides*). The adult female wasp usually infiltrates the host nest, lays her egg, and leaves, or destroys the egg of the host before she lays her own. The kleptoparasitoid larva consumes the provisions left for the host by its mother in the absence of the host larva, starves the host larva, or kills the host larva and then proceeds to dine. This may be one route to true parasitoidism. The next step is waiting until the host larva finishes eating, and then attacking it as an external parasitoid.

A strange form of parasitism afflicts certain social wasps. There are species that usurp the nests of other species and use the resident workers to raise their offspring. This is "brood parasitism" or "social parasitism." Several species are facultative brood parasites that can exist as either free-living species or as social parasites. The paper wasp *Polistes nimphus* can establish its own colonies independently, or a female can sneak into a *Polistes dominula* colony and lay her eggs. Other species are built for hostile takeovers, with bigger heads, and larger jaws and forelegs, like *Polistes sulcifer*, *P. semenowi*, and *P. atrimandibularis*. Among yellowjackets, five known species are obligate social parasites incapable of producing their own worker caste. The invading queen kills the resident and forces the host workers to raise her daughter queens, and sons.

The Social Wasps

Hornets, yellowjackets, and paper wasps are truly social (eusocial) wasps we regard with disdain if not hatred. Our contempt for them, beyond their sting, likely stems from their ability to exploit our habits and habitats with great efficiency. They raid picnics and barbecues, and nest on and in our homes.

On Guard
A worker European hornet, *Vespa crabro*, peers out the entrance of her nest. Worker social wasps perform many functions for their colony.

A wasp colony is a matriarchal enterprise, governed by a single reproductive female through aggression. The antagonism serves to prevent the ovaries of subordinate females from developing. This is in contrast to ants and social bees where the queen(s) rule with pheromones that worker females respond to. Colonies vary in size depending on species and environmental factors. Stenogastrinae and Polistinae typically have small colonies, whereas yellowjackets living in climates where they achieve perennial colonies can number tens of thousands. All social wasps are predatory, at least in part, feeding their larvae animal protein. Feeding is not a one-way street, however. Larvae produce a nutritious fluid eagerly imbibed by the adults. This mutual feeding, called trophallaxis, helps strengthen colony bonds between adults, as well as between adults and larvae.

The nests of most social wasps are fabricated from dry plant matter that is chewed and, with the wasp's saliva, fashioned into paper. The quality of this material varies greatly. Some tropical species create incredibly durable "carton" nests. Since the combs are filled with helpless eggs, larvae, and pupae, the adult occupants vigorously defend their nests from potential threats. This tendency is not lost on other animals. In the neotropics, oropendola birds, which create hanging nests of their own, often choose trees occupied by social wasps, or nest in close proximity to the insects. The wasps help protect the birds from predators and ectoparasitic insects.

Beyond hunting for prey, feeding each other, tending larvae, and expanding and defending the nest, social wasps may engage in other important tasks to sustain the colony. Fanning their wings, some workers aid in cooling the nest, and after a deluge, wasps may remove water from the nest, droplet by droplet. Individual cells need to be cleaned, destroyed, or built, while nectar may need to be stored in cells not used for brood. There is generally some degree of division of labor, usually by age. Younger adults tend to be stay-at-home nestmates, while older workers take on more risky jobs like foraging. Males, produced with new queens only once during the colony life cycle, have no occupation other than to mate with reproductive females, ideally from other colonies.

There is much to admire about social wasps. Instead of destroying their nests, maybe human homeowners should take a cue from oropendolas and hang signs reading "Beware of Guard Wasps!"

Social Predators
Social wasps like this *Polistes dominula* worker do not store prey through mass provisioning of paralyzed hosts. Instead, they hunt, kill, and pulverize prey to feed directly to the larvae in the nest.

SPOTLIGHT ON

Nocturnal Wasps

When the sun sets, we expect wasps to retire to their nests, but there is a "night shift" of species that we are usually unaware of. These wasps have arrived at this lifestyle for a variety of reasons, and adapted in unique ways.

The compound eyes of nocturnal insects are fundamentally different from those of diurnal (day-active) insects. The ommatidia are structured in a way that amplifies incoming light, typically with larger lens facets and wider rhabdoms. Meanwhile, the ocelli or "simple eyes," arranged in a triad at the crown of the insect's head, are greatly enlarged—the better to orient to the horizon with limited illumination.

Nocturnal behavior is how many ectothermic (cold-blooded) animals cope with extreme heat. It is no surprise that many wasps living in deserts emerge after dark when temperatures are cooler. This is especially true of velvet ants. Most are at least crepuscular, confining their activities to dusk and dawn, but many are strictly nocturnal. The Chyphotidae and Rhopalosomatidae, relatives of the mutillids, also include nocturnal species.

Another reason for night activity is because that is when hosts of various parasitoids are most vulnerable. Many caterpillars in particular feed out in the open at night. Some Braconidae and Ichneumonidae that seek them are exclusively nocturnal. Turn on your porch light and you are likely to attract them. Members of the ichneumon genus *Netelia* are considered a nuisance because of their abundance, and the females can sting.

The moon wasp, *Apoica pallens*, is so named because it is most active during a full moon. This social species, and ten others in the genus, live in the tropics of Central and South America. Nightly foraging begins with an abrupt, explosive departure of hundreds of wasps from the nest. They quickly return, then head out more randomly.

"Night hornets" of three species in the genus *Provespa* live in rainforests of southeast Asia. Like *Apoica*, new colonies are formed by swarming emigration from existing nests, much like honey bees. These wasps, as well as the European hornet, *Vespa crabro*, are wattracted to lights at night. The European hornet enjoys no special modifications for night vision, but its large size may simply confer the benefit of bigger eyes.

Daylight Defense
Moon Wasps, *Apoica pallens*, protect their nest during the day by arranging their bodies as a living fortress over the vulnerable brood of eggs, larvae, and pupae. At sunset they will erupt suddenly from the nest to begin foraging for prey.

7

Behavior
Instinctive Complexity

Intricate Adaptations

Survival of the individual, and the species, hinges on the success of behavioral adaptations. Wasps have developed complex and unlikely strategies to ensure genes are passed from one generation to the next. Instincts are not as inflexible as one might expect—perhaps less so than our own rigid, unflattering opinions of wasps.

Finding a host for her offspring is the overriding concern of a female wasp. Her species may be a generalist, accepting of several possible hosts, or highly specific, where only one species will suffice. In either case she must deliver her egg(s) in a way that guarantees success, though she has little control over future hazards in the form of predators, other parasitoids, and diseases. She acts alone or, for social wasps, recruits her daughters as labor to care for their younger siblings. She may exercise a surprising degree of parental care herself, even as a single mother.

Male and female wasps must also feed themselves to generate the metabolism necessary for movement. Flight is expensive, as is fleeing a predator or chasing a potential mate. Flower nectar is the fuel of choice for many wasps, but that is not the only option. Sugary wastes excreted by other insects are an alternative, and so is sap oozing from plants. Social wasps have learned that our sweet human beverages will also serve them nicely.

Male wasps have a single mission: find a mate, preferably a virgin. Competition is keen and males answer the challenge in myriad ways. Some are territorial, driving away other suitors, while other species embrace their comrades long enough to form an aggregation that passing females easily notice. A few seek higher ground—the better to survey for the fairer sex—and in some instances, males behave like females to fool their rivals.

Adult wasps must also groom, rest, maintain optimal body temperature, and ward-off dangers to themselves and their nestmates. Thermoregulation is the art of staying comfortable when the world around you is too cold or too hot. Sleep is not a behavior we associate with insects, but wasps need rest. They may sleep alone or together. Avoiding confrontation with potential predators can take the form of acoustic behaviors that alert the enemy of their ability to defend themself.

The pinnacle of complexity in wasp behavior might be the construction of nests, fortresses designed to protect the vulnerable egg, larva, and pupa stages of the wasp life cycle. Social wasps create elaborate nests, often very large, but solitary wasps are no less capable of crafting exquisite nurseries of their own. In fact, the nests of wasps are often more conspicuous than the insects themselves, and have their own fascinating histories.

Victorious Hunter
A female *Sphex* sp. digger wasp flies back to her burrow with a paralyzed katydid nymph. Solitary stinging parasitoids are generally specialized to seek one particular kind of host.

Host-seeking Behaviors

What wasps lack in visual acuity, they more than compensate for in their ability to utilize chemical and tactile stimuli in finding hosts.

Tightrope
The caterpillar, sensing the approach of this female ichneumon wasp, repelled on a strand of silk. Undeterred, the wasp crawled down the line and is about to lay an egg in the caterpillar. Her larval offspring will be an endoparasitoid.

Plants betray their identity to herbivorous insects through unique volatile compounds; an aromatic fingerprint. Female sawflies use these cues to target the appropriate host plant for oviposition. Emerging evidence suggests that insect larvae may also recognize chemical cues, crawling to a different plant or even choosing a different host. A study of the sawfly *Mesoneura rufonota*, a pest of camphor, supports this theory.

The information conveyed in a wafting scent is staggering. Plant perfumes also attract some parasitoids to their insect hosts. The female ichneumon *Itoplectis conquisitor* attacks caterpillars found only on Scots pine, so responds to the fragrance of that specific tree. The odor of collard greens draws the parasitoid *Diaeretiella rapae*, where it seeks aphid hosts. Freshly damaged plants emit unique semiochemicals that alert parasitoids to the presence of their hosts.

Parasitoids that oviposit into insect eggs may search for eggs directly, using their antennae to fondle the ovum and determine if it is viable, the right size, shape, and/or texture. Others ride an adult female of the host (phoresy) until she lays her eggs. Egg and pupa parasitoids may leave a scent designed to deter a competing parasitoid from usurping the same resource.

Bloodhounds have nothing on wasps like the braconid *Cardiochiles nigriceps*, which sniffs out the residue left by its caterpillar host's salivary secretions. Still other parasitoids track hosts through traces in frass (insect feces), or silk strands left by caterpillars.

Some parasitoids do rely on vision locate a host. Beewolves in the genus *Philanthus* fly slowly around flowers in search of bees. Mud dauber wasps in the genus *Chalybion* notice the webs of their spider hosts while flying. They then land on the web and deftly pluck the snare to mimic a struggling insect. This draws the spider to its doom. Mason wasps (Vespidae: Eumeninae) that specialize on concealed caterpillars recognize the leaf rolls and "sandwiches" made by their hosts. Unable to breach the leaf barrier, the wasp antagonizes the occupant until it ejects from an opening, whereupon the wasp seizes it.

Rather than confronting their stinging wasp hosts in the hosts' nests, some cuckoo wasps in the genera *Pseudolopyga*, *Holopyga*, *Pseudomalus*, and *Omalus* insert their eggs into the living true bug hosts those stinging wasps are seeking. The host wasp must take that particular insect to its nest in order for the cuckoo's offspring to succeed as a cleptoparasite of the host wasp.

Predatory and scavenging social wasps use both visual and chemical clues to maximize foraging success.

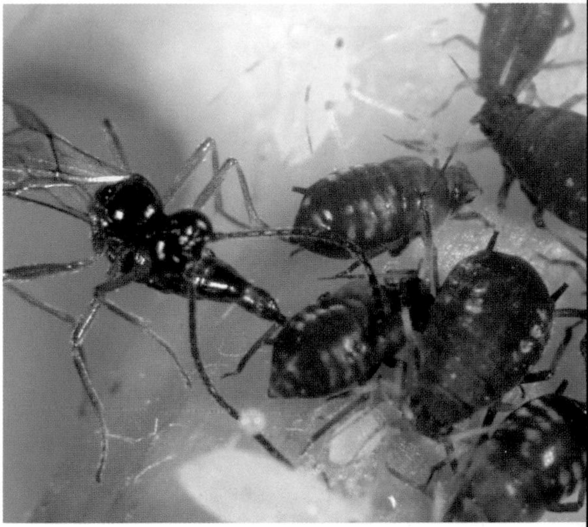

Persistence
The forest of long hairs on this Indian caterpillar (below) are designed to discourage parasitoids, but this ichneumon wasp is undeterred as she injects an egg inside it.

Mummy Maker
A female braconid, *Aphidius ervi* (above), injects an egg into a black bean aphid host. Her larval offspring will feed inside, bloating it into an "aphid mummy" before emerging as an adult wasp.

SPOTLIGHT ON

Thermoregulation

Wasps have a narrow range of temperatures in which their bodies function at optimal capacity. They cope with excessive heat and extreme cold in some ingenious ways. Increasingly, scientists are finding these insects to be less at the mercy of the elements than previously thought.

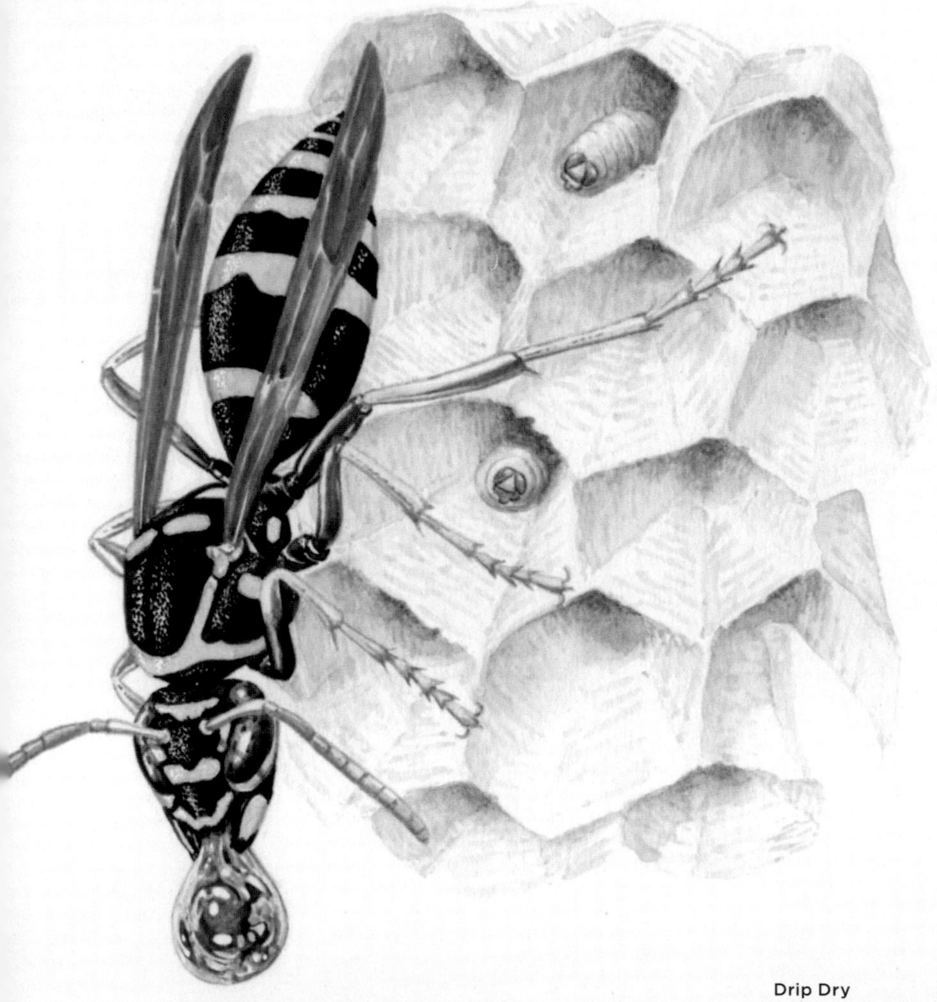

Drip Dry
A paper wasp, *Polistes* sp., releases a drop of water she gleaned from the nest after a rain shower. A nest in danger of overheating may be cooled by applying water droplets.

Paper Palace
The multilayered paper envelope (right) of a yellowjacket nest helps insulate the vulnerable occupants from excessive cold in the northern climates where these wasps live.

Heat Shield
The short, reflective silver hairs on this female thread-waisted wasp, *Ammophila ferrugineipes* (below, right), protect her from overheating in the desert sun of South Africa as she digs her burrow and hunts for caterpillars.

Species of wasps living in deserts and other broiling habitats are often built to take it. They are frequently pale in color, and/or coated in reflective hairs or scales of silver or gold. Where structural modifications end, behavioral strategies take over. Wasps avoid scorching by standing on tiptoe, or perching frequently on vegetation just off the ground where the temperature can be profoundly cooler. It may be more comfortable below the surface, too, so some wasps dig beneath the sand to endure hot spells. A surprising number of wasps are nocturnal.

The larvae of the Australian sawfly *Perga dorsalis* adopt a posture called "obelisking" when things get too hot. By orienting their bodies as vertical as possible at midday, they minimize the body surface exposed to direct sun. Hotter still? The larvae then resort to coating their bellies with anal secretions that cool them through evaporation.

Some wasps may not become active on cool, cloudy days, but even morning hours can be chilly. Shivering wing muscles can elevate their body temperature above the air temperature, effectively making the insect "warm blooded." Many wasps bask early in the day, sprawling and pressing their bodies against a stone or other warm substrate. They warm by absorbing heat from the ground (conduction), as well as through convection via the warming air immediately above them.

Regulating the climate of a nest is a challenge met with architectural engineering and behavior. Yellowjackets build paper combs enveloped in a multi-layered paper covering that keeps the brood warm. The entrance to the nest is small, and at the bottom to limit heat loss. Young nests may have the entrance elongated into a funnel, perhaps enhancing heat conservation even more, earlier in the season when temperatures are cooler. The exposed combs of the mostly tropical paper wasps can require cooling. The adult wasps may fan their wings as living air-conditioning units. Failing that, water may be carried to the surface of the nest to cool it through evaporation.

Adult Feeding Behavior

Adult wasps need fast food to sustain their frenetic pace. Like any human athlete, carbohydrates are what wasps crave, and they find those sugars in a variety of places.

Drink Up
Most wasps need to drink water, especially paper wasps like this *Polistes* sp. that need to manufacture saliva to create the wood pulp they use to fabricate their nests.

Flower nectar is a rich source of fuel, so many wasps visit blossoms frequently. They prefer composites, umbelliferous flowers, and floral racemes that represent numerous blossoms in a group, with plenty of visibility so they can detect approaching predators or parasitoids. Few wasps dive headlong into tubular flowers.

Many plants do not limit their nectar output to flowers. Extrafloral nectaries may exist on any number of plant parts, including bracts, leaf petioles, stems, stipules, cotyledons, fruits, and pods. Thus, even when a plant is not blooming it may still produce substances attractive to wasps. It is another tactic plants use to attract allies in their war against herbivorous pests.

Sap from injured plants oozes, then ferments, drawing many insects, including wasps. In Southwest U.S.A., desert broom (*Baccharis* spp.) is well known for this phenomenon. Tarantula hawks, cicada killers, the steel blue cricket hunter (*Chlorion aerarium*), and other wasps flock to the "bleeding" shrubs. Wasps may not always wait for a tree injury. The European hornet, *Vespa crabro*, actively girdles the stems of lilac and other shrubs to start a sap flow.

Honeydew is the sugary liquid waste of aphids, scale insects, and other true bugs that siphon plant sap. The insects must draw copious amounts of sap to derive minimal nutritional value, and they expel an inordinate amount of waste in the process. Wasps, bees, flies, beetles, moths, butterflies, and others take full advantage of the output. Ants jealously guard aphid colonies from competing honeydew-seekers. Honeydew is a critical food source at times when few flowers are in bloom, especially in late fall.

Food scarcity can lead to horrific solutions. Many social wasps, such as the Moon Wasp, *Apoica pallens*, will cannibalize brood (eggs, larvae, pupae). This brutal practice feeds the adult wasps and reduces the number of larvae they need to nourish.

Hornets and yellowjackets frequently kill bees to eat the honey stored in a bee's crop, but other adult wasps feed on invertebrates, too. Ironically, *Tenthredo* sawflies that eat plants as larvae will prey on insects as adults. Pompilid spider wasps routinely nibble on spider hosts before caching them as food for their offspring. In the tribe Ageniellini, many species amputate the legs of the spider to make for easier transport. The wasps lap up fluids from the resulting wounds. Tiny scelionine wasps, *Mantibaria*, live as blood-sucking ectoparasites of large mantises while waiting for the female mantis to deposit eggs that the female *Mantibaria* will then parasitize.

Refreshing Nibble
Many stinging parasitoid wasps, like this spider wasp, *Auplopus* sp., will sip oozing body fluids from prey that is intended to feed their larval offspring.

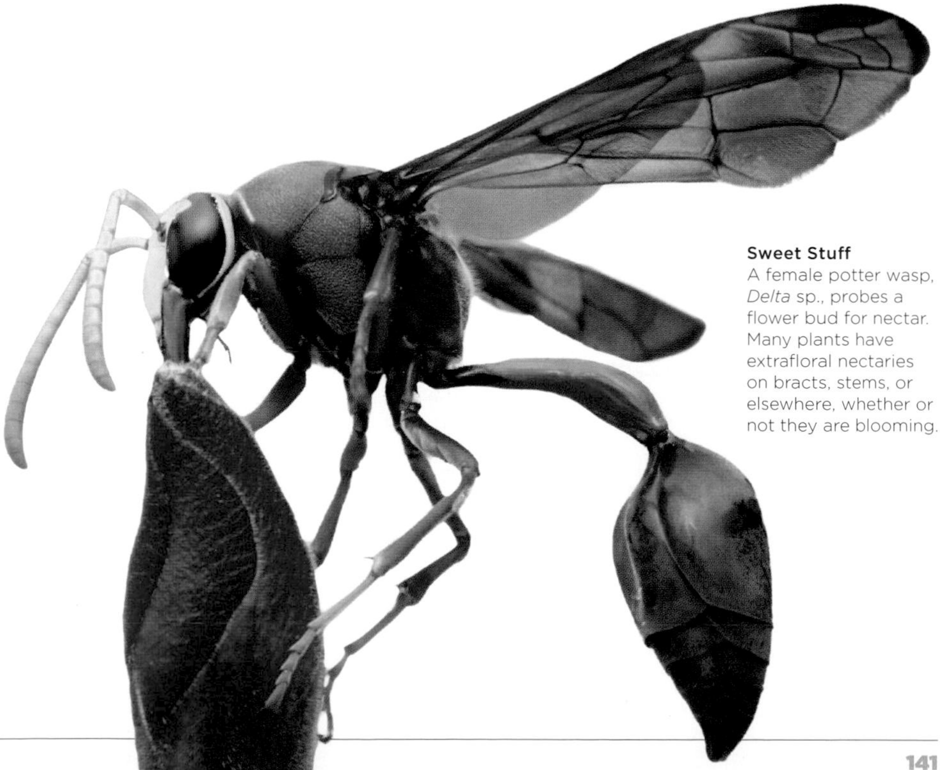

Sweet Stuff
A female potter wasp, *Delta* sp., probes a flower bud for nectar. Many plants have extrafloral nectaries on bracts, stems, or elsewhere, whether or not they are blooming.

SPOTLIGHT ON

Parental Care

The motherly instincts of wasps range from near total disregard to 24/7 devotion. Fatherly behavior, aside from fertilization of the female, is almost non-existent.

Protective Mother
A female sawfly, *Pseudoperga lewisii*, stands guard over her newly-hatched larvae in Australia. Such extended parental care is rare among solitary wasps.

Even sawflies will offer protection to their brood. No less than ten genera, across the families Argidae, Pergidae, Diprionidae, and Tenthredinidae have been documented to provide maternal guardianship to their eggs and young larvae. The warning colors of female Brazilian argids, *Themos olfersii* and *Dielocerus diasi*, may help discourage visual predators. *T. olfersii* sprawls across her clutch, on a leaf, while *D. diasi* sits at the base of the leaf where her eggs are more widely dispersed. The approach of an ant, egg parasitoid, or other threat may be met with a menacing posture, wing buzzing, lunging, biting, mantling over the eggs, or a "push-up" display.

Among stinging parasitoids, many members of the family Bethylidae provide sentinel services to their young. *Sclerodermus harmandi* is a parasitoid of wood-boring larvae of the beetle *Monochamus alternatus* in Asia. Females tend their small clutch of eggs until they hatch. *Goniozus nephantidis* females remain with their offspring until they pupate. *Prosierola bicarinata* of Mexico protects even the pupae of her brood.

Some female sand wasps in the genera *Bembix*, *Stictia*, and *Rubrica* practice progressive provisioning, feeding their larvae as needed. A single wasp maintains not only several cells per nest, but often multiple nest burrows. Mom can tailor the prey she provides to the developmental stage of each larva. She visits more frequently than a wasp that caches prey all at once, so she is better able to detect parasitoids and inquilines. Females of the enormous Australian potter wasps of the genus *Abispa* foster one larva at a time inside a mud cell they guard ferociously. Paralyzed caterpillars are furnished in a manner described as "truncated progressive provisioning," before the cell is finally sealed and another one constructed adjacent to it.

Social wasps employ sisters to care for younger siblings, resulting in a greater number of individuals with around-the-clock supervision. It is the ultimate in quantitative daycare. In at least three species of *Polistes* paper wasps, and ten species of Vespidae total, males have been observed to feed their larval sisters. This is brotherly care, not fatherly behavior.

Squadron
Several female sand wasps, *Bembix* sp. (below), each bring flies to feed their larvae. They are solitary, each digging her own burrow, but often nest close together in dunes, on beaches, and other expansive sand habitats.

On Demand
Female sand wasps in the genus *Bembix* (right) practice progressive provisioning: they bring flies to their larvae as needed, instead of caching prey all at once. Some *Pemphredon* spp. aphid wasps do likewise.

Acoustic Behavior

Wasps employ an array of sound effects, including vibration, for a variety of purposes. Acoustic signals are used in locating hosts, courtship, self-defense, colony defense, displays of dominance, and food solicitation.

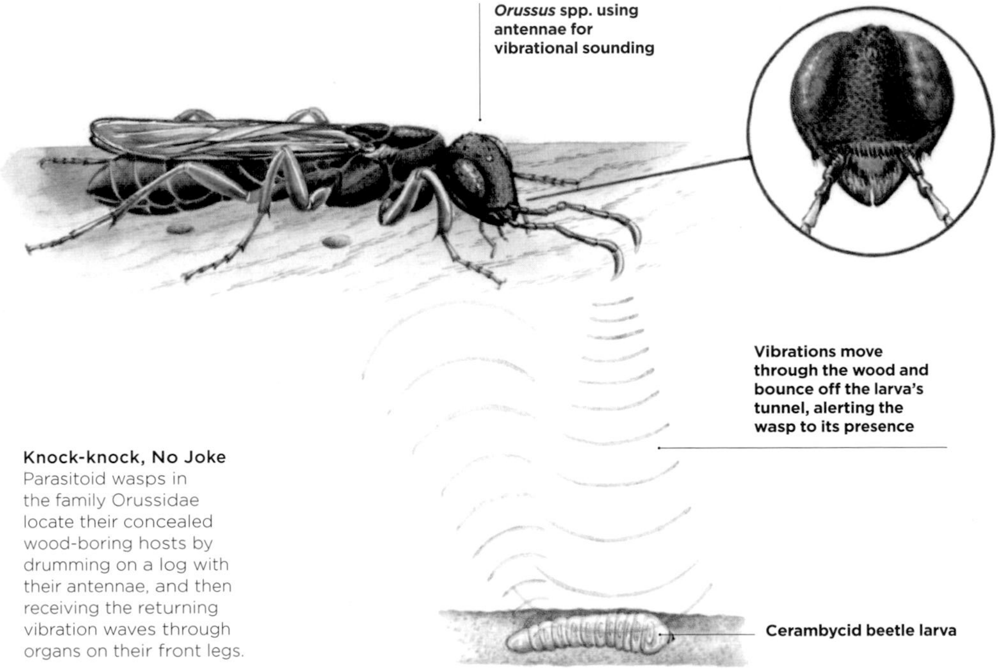

Orussus spp. using antennae for vibrational sounding
Vibrations move through the wood and bounce off the larva's tunnel, alerting the wasp to its presence
Cerambycid beetle larva

Knock-knock, No Joke
Parasitoid wasps in the family Orussidae locate their concealed wood-boring hosts by drumming on a log with their antennae, and then receiving the returning vibration waves through organs on their front legs.

Parasitoid wasps specializing on concealed hosts face the challenge of finding them within a dense substrate. Many species use vibrational sounding, akin to bat echolocation, to detect a host. Female *Orussus* spp. have modified antennae for tapping on logs and tree trunks, generating shock waves through the wood. Enlarged hearing organs on their front legs receive the reverberations as the shock waves return. Differences in returning waves can indicate a void, such as the tunnel of a host, or reveal the presence of the host itself. Many ichneumons also use vibrational sounding.

Wing-fanning is typical of many male parasitoid wasps during courtship. It serves two functions: one is to waft pheromones toward a female; the other is to produce a "song" to woo her. Both are messages of male fitness for her evaluation. Females of the braconid aphid parasitoid *Lysiphlebus testaceipes* prefer pulses of higher frequency, and higher wing amplitude during fanning. The song of the microgastrine braconid *Glyptapanteles flavicoxis* is more complex, with low-amplitude buzzes alternating with a higher-amplitude, higher-frequency display.

Many wasps produce buzzing from wing muscles. It burns a lot of energy, and raises body temperature, but can assist in activities such as digging a burrow in dense soil. In the case of the eastern cicada killer, *Sphecius speciosus*, buzzing may ward off small vertebrate predators, as well as deter competing females from usurping a nest burrow. Buzzing creates vibration, too. Female social wasps *Parischnogaster nigricans*, incorporate wing buzzing into threat displays to maintain social dominance. Female *Eusteonogaster calyptodoma*, primitively social stenogastrine wasps, rattle their abdomens against the walls of their nests to deter attempts by other females to usurp a nest.

Velvet ants in the family Mutillidae generate sound by stridulation (see page 82) when in distress. Rubbing abdominal segments together requires less energy than wing buzzing, and is still clearly audible. In at least two species of "panda ants" in Chile, *Euspinolia chilensis* and *Euspinolia militaris*, there is an ultrasonic component to their auditory display. One theory suggests this repels certain predatory rodents.

Sound-making and vibrational cues are not limited to adult wasps. The larvae of many social wasps rasp their mandibles against the walls of their paper cells to solicit food from adults. *Polistes* paper wasps are known to drum their antennae on the edge of a cell to alert the larva within that it is about to receive food.

Turn the page for the most intimidating wasps sound of all.

Can You Hear Me?
Most velvet ants, like this female "panda ant," *Euspinolia militaris* of Chile, squeak audibly in response to threats. This species is unique in producing an ultrasonic frequency as well.

PROFILE

Warrior Wasps

Synoeca spp.

To say that the six species of social wasps in the genus *Synoeca* live up to their reputation of being unbelievably fierce is an understatement. but you cannot say they do not give fair warning. They are best known for their unique nest architecture and hair-raising mass threat display.

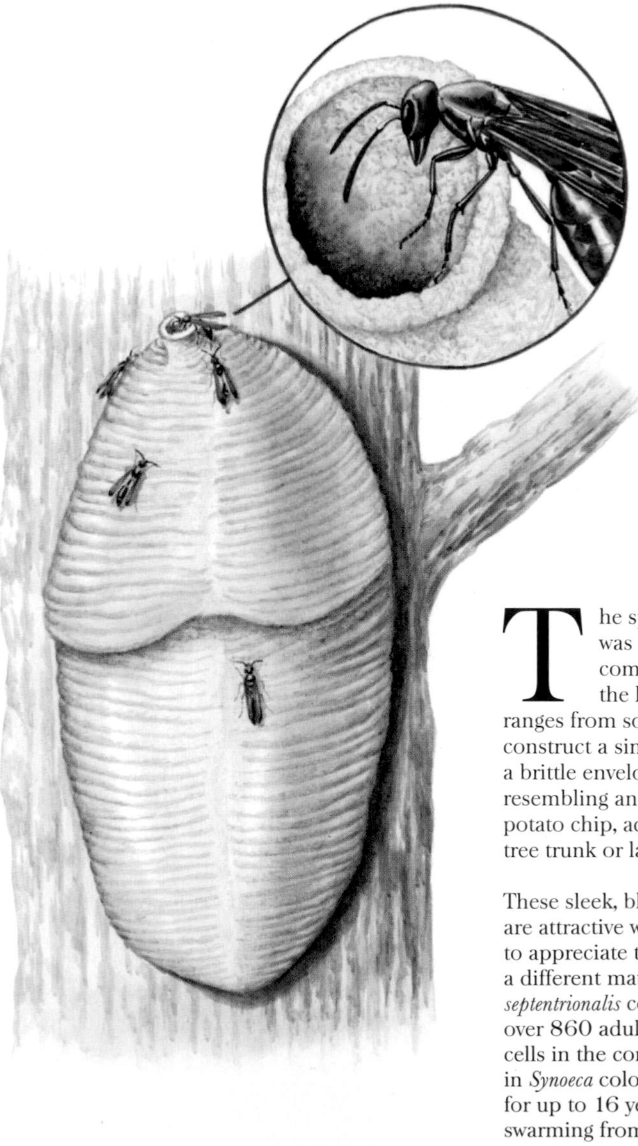

Full of Fury
Tropical habitats are full of danger, so social warrior wasps of the genus *Synoeca* have evolved intimidating strategies to protect their vulnerable nests. This includes a loud warning display, directly followed by assault.

The species *Synoeca septentrionalis* was only described in 2017, but its compatriots have been known since the late 1700s. Collectively, the genus ranges from southern Mexico to Argentina. They construct a single-layer paper comb covered in a brittle envelope that has been described as resembling an armadillo or a gigantic, ribbed potato chip, adhered flush against a smooth tree trunk or large branch.

These sleek, blue-black insects with black wings are attractive wasps if you dare get close enough to appreciate their beauty. An entire colony is a different matter. The largest nest of *Synoeca septentrionalis* collected in one study contained over 860 adult wasps, and more than 1,400 cells in the comb. There are multiple queens in *Synoeca* colonies, and one nest can persist for up to 16 years. New colonies are formed by swarming from existing colonies.

Genus *Synoeca*

SPECIES	6
DISTRIBUTION	Southern Mexico to Argentina
SIZE	~1 inch (~25mm)
AMAZING FACT	*Synoeca ilheensis* was only discovered in 2017

You will be made aware of a nearby colony of warrior wasps in no uncertain terms. The occupants react to an approaching threat by drumming loudly, synchronously, and rhythmically against the interior surface of the nest envelope. Fail to heed the warning and the occupants will pour out of the nest entrance, spill over the exterior, and continue drumming. "Can you hear us now?!" Shortly thereafter they commence the assault. Their stings are barbed, like those of honey bees, so while they sting repeatedly, eventually they can become hopelessly anchored in the flesh of a victim and die defending their nest.

Justin O. Schmidt, creator of the Schmidt Sting Pain Index, gives *Synoeca septentrionalis* his highest rating, a full four on the zero-to-four scale. He writes of the sting from just one individual: "Torture. You are chained in the flow of an active volcano. Why did I start this list?" On a more positive note, a team of Brazilian research scientists discovered a new chemical compound in the venom that shows promise in treating anxiety. It has at least demonstrated great potential in trials with rats.

Actual size

Winged Warrior
Few social wasps are as aggressive in defense of their nests as *Synoeca* spp. It pays to heed their audible warning.

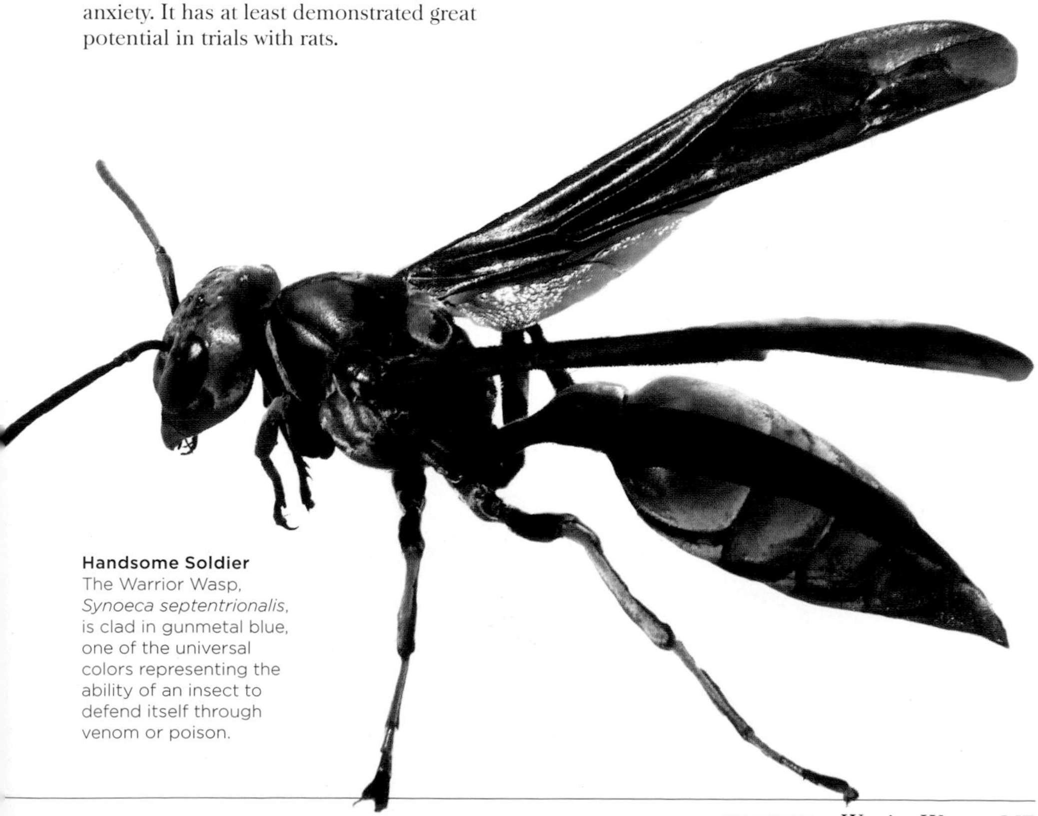

Handsome Soldier
The Warrior Wasp, *Synoeca septentrionalis*, is clad in gunmetal blue, one of the universal colors representing the ability of an insect to defend itself through venom or poison.

Sleeping

Wasp sleep is not equivalent to that of vertebrates, but does represent a prolonged period of daily inactivity. Wasps adopt some odd behaviors in "slumbering," and are less responsive to external stimuli.

Sleep for day-active wasps is closely tied to their ability to function physically at the lower temperatures of night, and to see in the absence of light during hours of darkness. Most wasps do not possess the ability to elevate their body temperature above the air temperature, at least not without metabolic penalties. Their eyes are also not adapted to dim light. Lastly, the hosts of parasitoid wasps may be concealed and inactive themselves, in which case pursuit of them would be futile.

Logic would suggest that it is best for a wasp to be as invulnerable as possible during a time when it is less responsive to external stimuli, as is the definition of "sleep" for wasps. Indeed, some wasps spend the night within nest burrows or, in the case of males, shallow sleeping tunnels they excavate for that sole purpose. A surprising proportion of other wasp species are conspicuously exposed during the night.

The normally solitary blue mud dauber, *Chalybion californicum*, forms dense aggregations of dozens, even hundreds, of individuals under eaves, rock ledges, and other sheltered locations at night. The sand wasp, *Steniolia obliqua*, also solitary in every other regard, gathers in sleeping "balls" on terminal branches of pine trees, and other vegetation. Other species, like the thynnid wasp *Myzinum quinquecinctum*, avoid bodily

Sleep Like a Stick
This ant-mimicking thread-waisted wasp, *Ammophila wrightii*, grips a twig and props itself at an angle. It may be joined by others on nearby vegetation in a loose group.

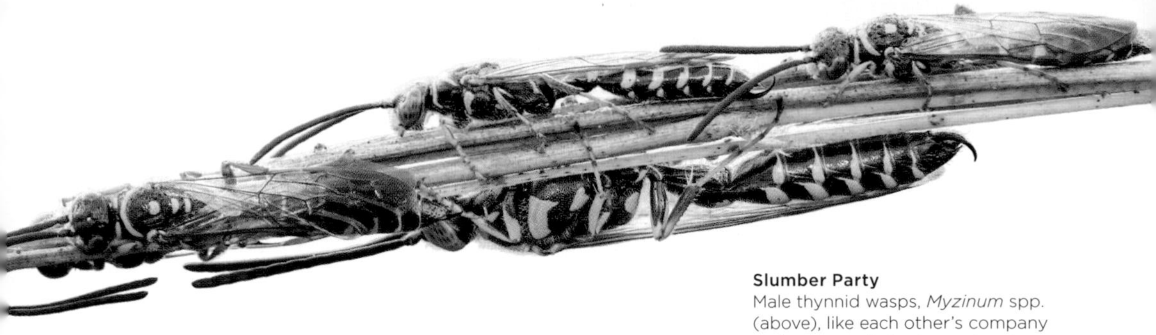

Slumber Party
Male thynnid wasps, *Myzinum* spp. (above), like each other's company and congregate in dense sleeping clusters on vegetation in open habitats. Never mind the menacing pseudosting, they are harmless.

contact, but still assemble in loose groups of many individuals—in their case, exclusively males. Grouping might reinforce aposematic colors during daylight, but at night? Why some wasps hold "bachelor parties" while others are not segregated by sex is also a mystery. It could be that it facilitates mate-finding behavior come daybreak. Indeed, mating behavior has been observed in *Steniolia* clusters.

Search tall grasses around dusk and you may find thread-waisted wasps of the genus *Ammophila* beginning to bed down. They enjoy each other's company, but will jostle each other for the most favored perches. Once satisfied, each wasp clenches a stem in its jaws and props its body at an angle of 50–70 degrees. Nearby shrubs and trees may hold individual scoliid wasps like *Scolia dubia*, all of which are males, each curled in comfort around a flower of stem. Mason wasps (Vespidae: Eumeninae), and many cuckoo wasps (Chrysididae) also sleep curled around a stalk or stem. In one localized area it is possible to see dozens of unrelated wasp species, and also solitary bee species, engaged in sleep behavior. This is especially true in prairies and deserts.

Sleep Tight
A female solitary thread-waisted wasp, *Prionyx* sp. (above), clamps onto a stem with her jaws in preparation for an overnight sleep.

Curled Up
A solitary mason wasp (left) sleeps by locking her jaws onto the tip of a leaf and curling her body around it. This is a typical snoozing posture for eumenine vespids.

Male Behaviors

Males are the expendable sex in the animal kingdom, and they act like it, too, behaving with reckless abandon. Wasps, anyway. It was not until the 1960s that scientists began researching male wasps in earnest. Today, entire books could be written on the subject.

Keep Out!
A male keyhole wasp, *Trypoxylon* sp., defends his mate's tubular nest from an advancing cuckoo wasp. Male cooperation in nesting is not common among wasps, be they solitary or social.

Lekking
Male horntail wasps, like the *Urocerus gigas* shown here, gather at the tops of isolated or prominent trees where they are likely to be seen by passing females. This behavior is known as lekking.

The boys are interested in knowing where the girls are. Mate-seeking in male parasitoid wasps is similar to host-seeking in females: They respond to olfactory, tactile, and sometimes acoustic stimuli. Males typically emerge earlier than females, and immediately seek a virgin mate. Male giant ichneumons, *Megarhyssa* spp., detect the sound of another wasp gnawing its way to freedom from inside a log or tree. Several males of different species may gather in anticipation, but it might be another male or the wrong species about to crawl out. Most successful mating occurs before a female emerges. The long, slender male abdomen extends well into her tunnel to reach her.

Stinging parasitoids may or may not be more sophisticated. Timing, receptivity of females, location, ability to demonstrate individual fitness, and competition with other males, drives male behavior. Individual nests in close proximity result in territorial males that attempt to mate with multiple females, as in cicada killers (*Sphecius* spp.). Males of cavity nesting species may end up mating with their sisters, as in some mason wasps (Vespidae: Eumeninae).

Where nests are not concentrated, males adopt other solutions. They may patrol flower patches and apprehend visiting females, or stake out another resource, such as a water hole where females gather mud (some pollen wasps, mason wasps). Males of some *Mellinus* spp. frequent fresh animal feces where the female wasps hunt flies.

Males of other species use landmarks, defending isolated perches, ridgetops ("hilltopping"), and other promontories where the likelihood of a rendezvous with a female is enhanced. The tarantula hawk *Hemipepsis ustulata* of the southwest U.S.A. is a well-studied example. One male held territory for 40 consecutive days. Some pollen wasps (*Pseudomasaris* spp.) are also hilltoppers. Male stink bug hunters, *Astata* spp., possess enormous eyes. They perch atop a tall weed and dart out in lightning-quick flights to pursue a female or evict a competing male.

Hilltopping and landmark use is a subset of lekking behavior in which groups of males display and/or congregate. Besides the tarantula hawks, paper wasps of the genus *Polistes* gather around tall objects in the landscape. Male horntail wasps (Siricidae) release special scents called aggregation pheromones that draw both males and females.

Some beewolves in the genus *Philanthus* lek also, but all attract females by scent-marking. Other wasps scent mark, too, including *Polistes dominula*, as part of landmark territory defense.

Courtship and Mating

Romance is sadly lacking among wasps. Male competition often occurs until the moment of consummation, leaving little time for sentimentality. Still, wasp mating systems are overwhelmingly based on female choice, so males have something to prove.

Stacked up
Four male sand wasps, *Bembix* sp., have piled onto a single female in a strategy called "scramble competition." The frenzy will end with one winner, but individual wasps may suffer injury in the melee.

Males of some wasp species impress with a song-and-dance routine. The male pteromalid *Pteromalus coloradensis* rocks side-to-side and fans his wings. Courtship of the pteromalid *Nasonia vitripennis* involves sound, smell, and touch. Aging males produce less pheromone, so must up their game in the talent portion. The chalcid *Brachymeria intermedia* has a three-part song medley. Stenogastrine social wasps earn the moniker "hover wasps" for the aerial displays of some species. Males posture for each other, females, or both, showcasing white abdominal bands.

Wingless female Australian thynnid wasps face the challenge of getting to nectar or other carbohydrate foods, so they summon winged males with a pheromone they emit. Depending on the species, the male may bring her food in a droplet under his "chin," regurgitate food, excrete food from his anus, or carry the female to sources where she serves herself. Flying with a female in tow is called phoretic copulation. This is one rare instance where males are larger than females, and where males invest enough to have their choice of partners. Phoretic copulation is also known in velvet ants (*Timulla* and other genera), and bethylid wasps (*Apenesia nitida, Dissomphalus* spp.).

Intimate contact with the antennae and head of the female is often necessary for mating success. Male shield-handed wasps, *Crabro* spp., have translucent flanges on the front tibiae. The pattern is species-specific and, when the male alights on a potential mate, he covers her eyes with these plates, filtering light in a manner that may make her receptive to mating. A female pteromalid wasp in the genus *Muscidifurax* demands proper fondling before shape-shifting her abdomen to allow the male access.

The ultimate trick might be played by male *Cotesia rubecula* braconids. Mated males, while still atop their mates, mimic the submissive posture of a receptive female. It fools and distracts rival males from his mate, since she could mate yet again.

Phoretic Copulation
A male velvet ant (above, winged) has flown off with his smaller, wingless mate. Larger males carrying smaller females while engaged in mating is common behavior in Mutillidae, Thynnidae, and some Tiphiidae.

After mating, males rarely hang around, but it pays to do so if the female is likely to mate again with a different male. Mate guarding is one response. The thread-waisted wasp *Eremnophila aureonotata* is frequently seen in tandem, on flowers and in flight, male atop female. A male organ pipe mud dauber, *Trypoxylon* spp., goes farther still, guarding the nest against parasitoid flies and other enemies while his mate is foraging. He may also assist in storing prey, but he expects another sexual encounter each time she returns.

Close To You
Male (left, top) and female (left, bottom) thread-waisted wasps, *Eremnophila aureonotata* of North America, remain engaged as the female feeds on nectar. The tandem can even fly together seamlessly.

Nesting

Building nests is an extension of parental care. Nests serve as nurseries for helpless offspring, food storage containers, climate-controlled chambers, and sometimes social structures.

Many wasps simply deposit eggs on or in a host and provide no further care; galls are nests of sorts, but the wasp makes the plant do the work. A pre-existing natural cavity where a paralyzed host can be sheltered out of reach of other organisms, and shielded from weather, may suffice as a nest for many stinging parasitoids. Modifications of those nooks and crannies were likely one evolutionary path towards true nest-building.

Underground burrows of fossorial (digging) wasps may be shallow or deep, simple or complex. Some build turrets at the entrance, while others dig fake "accessory burrows" nearby. Both tactics help thwart parasitoid wasps and flies from finding or entering the nest. Individuals may nest in aggregations to lessen the chance that *their* nest will fall victim to parasitoids, but they risk theft of prey and usurpation by their neighbors. Some species increase security by closing their nests while out foraging. The wasp makes a spiral orientation flight to memorize the location. Meanwhile, we forgot where we parked.

Going Underground
Most solitary stinging parasitoids dig subterranean burrows with multiple chambers. Sand wasps like *Bembix rostrata* (left) favor fine soil to dig into.

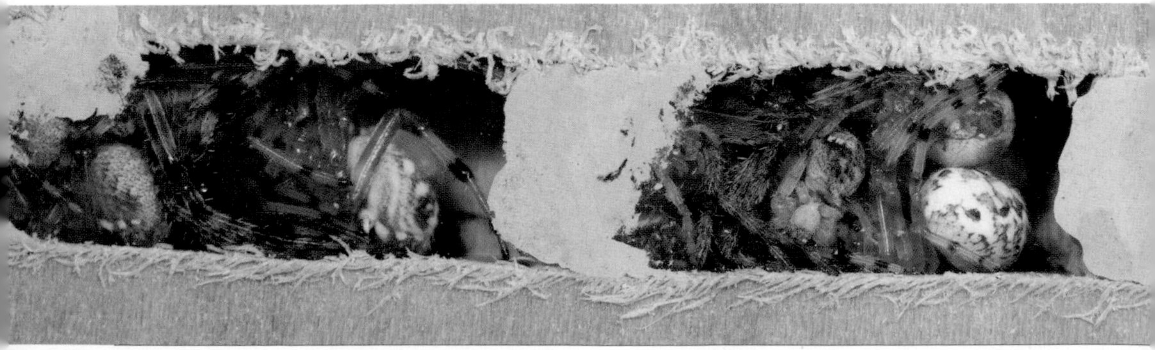

Shockingly Social
Females of the mason wasp *Montezumia cortesioides* border on being social. This nest in Costa Rica shows how several females may share a nest, but there is no division of labor.

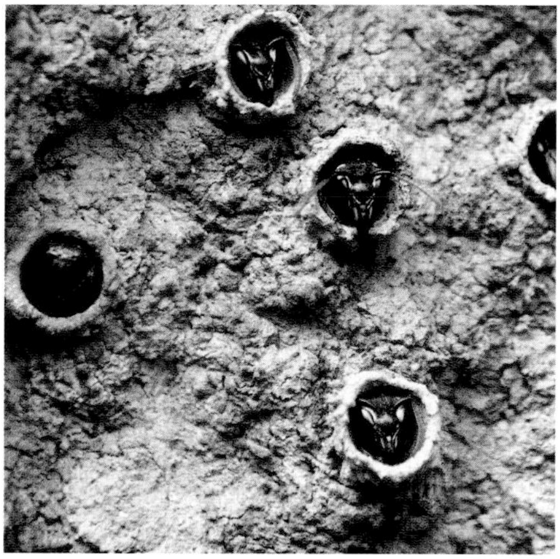

Hanging Home
This sparse, vertical nest is the work of a paper wasp, *Mischocyttarus* sp. (right), in the Kanuku Mountains of Guyana. Though delicate, it is cryptic and out of reach of most predators and parasitoids.

Wasps utilizing hollow plant stems, abandoned tunnels created by wood-boring insects, holes where nails or screws have worked loose, and other tubular cavities, leave an empty "vestibular cell" to fool parasitoids. Behind the vacant cell is a linear arrangement of one or more cells with partitions between them.

Free-standing mud nests are created by some members of the families Pompilidae (spider wasps), Sphecidae (thread-waisted wasps, mud daubers), Crabronidae, and Vespidae (pollen wasps, potter wasps, and mason wasps). Vessels may be singular or grouped, and represent the work of one or more individual females.

Linear Layout
A trap nest reveals how some solitary wasps, like this keyhole wasp, *Trypoxylon* sp. (left), divide a pre-existing cavity into chamber "apartments," one for each larval offspring. Their food is a cache of paralyzed spiders.

Social wasps achieve the zenith of size, durability, and diversity in Hymenoptera nests. Most are made of a paper-like material called carton: wood or plant fibers chewed into a saliva-rich pulp and molded into cells and often an envelope. Perennial nests of yellowjackets can reach enormous proportions. Nests have been found completely filling the interiors of derelict vehicles. *Microstigmus* species in the family Crabronidae are the only known social wasps outside of the Vespidae. Females work singly or cooperate in groups to fashion a tiny nest of fibers from the fronds of certain palms, together with silk manufactured in the wasp's abdominal glands.

Strangest of all are the "bubble" nests of certain wasps in the genus *Protopolybia* (formerly *Pseudochartergus*) in Amazonian rainforests, and *Ropalidia opifex* of tropical Asia. The insects wrap their nests in "wax paper," an opaque shield composed mostly of secretions from their mandibular glands, minus the usual plant fiber particles.

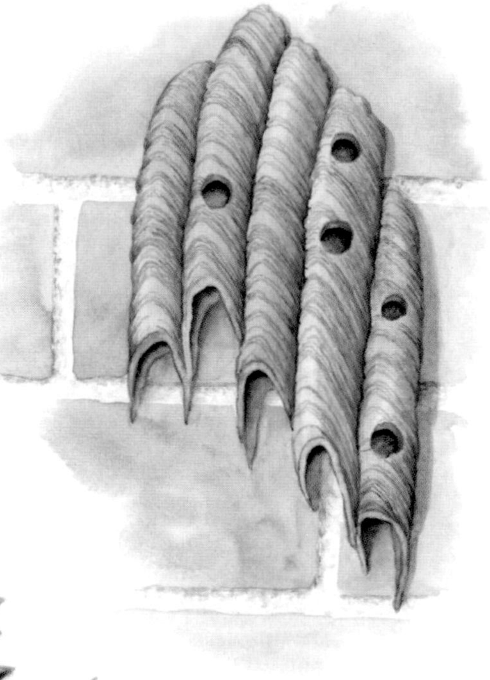

Tubular
Females of the organ pipe mud dauber, *Trypoxylon politum* of North America, are truly solitary, each female building her own nest. Each "pipe" contains several cells. Here, wasps have already emerged, leaving exit holes.

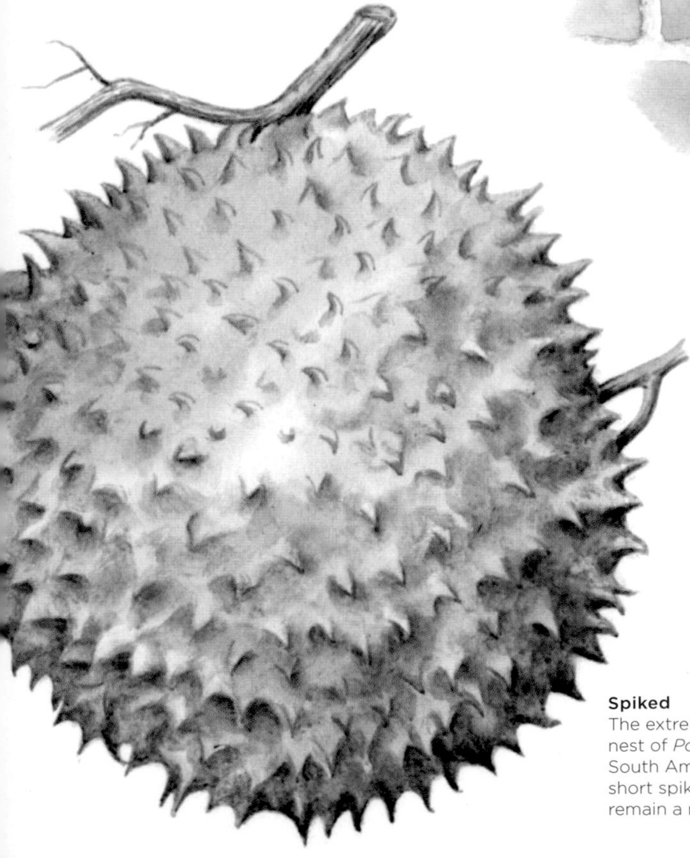

Spiked
The extremely durable carton nest of *Polybia scutellaris* of South America is covered in short spikes, for reasons that remain a mystery.

Enormous
The scale of some nests of tropical wasps, like this colony of a *Chartergus* sp. in Brazil, is difficult to convey. Year round warmth allows colonies to become perennial, with mansion-size abodes to accommodate them.

Fluted Masterpiece
The graceful architecture of this nest is the work of one or two semi-social wasps, *Eustenogaster calyptodoma* of Thailand, Malaysia, and Indonesia. The shape likely sheds rain during jungle downpours.

Vertical comb
The pedicel from which this nest of paper wasps, *Polistes instabilis*, is suspended is coated with a black substance that repels ants. the major enemies of social wasps.

8

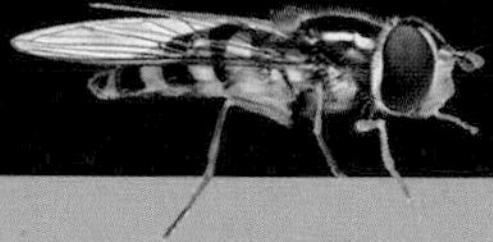

Wasp Mimics
Protection by Deception

How to Fool Your Foes

Stinging wasps are the evolutionary envy of the rest of the insect world. They have the ability to defend themselves with venom, and advertise the fact with aposematic "warning" colors. The common color schemes shared by stinging wasps have evolved through Mullerian mimicry, a phenomenon named in honor of German biologist Fritz Müller, who observed that most stinging or toxic insects resemble each other. English naturalist Henry Walter Bates noticed that perfectly defenseless organisms often look like dangerous ones. Sawflies and other insects not equipped with a sting can fake it by simply looking and acting the part, an evolutionary survival strategy called Batesian mimicry.

Both types of mimicry are aimed chiefly at visual predators, such as birds, lizards, and primates, but perhaps jumping spiders, too, which have exceptional eyesight. A more recent theory suggests another target of mimics: wasps themselves. Most insects are victimized in one way or another (usually several) by wasps, so why not look like a wasp yourself? Mimicry normally requires the sacrifice of at least one stinging model for a generalist predator to learn its lesson and apply the new knowledge, but few wasps are generalists.

Almost all insect orders count wasp mimics among their membership. Flies have nearly perfected mimicry of stinging wasps, but insects as unlikely as beetles, true bugs, moths, katydids, and mantidflies offer astounding examples that can cause even the most experienced entomologist to do a doubletake.

Frequenting the same habitats and practicing the same habits as your stinging "model" is the best way for a mimic to convince a doubting predator that it is concealing a venomous weapon. The sting is a retractable part of wasp anatomy, so even those species equipped with one rarely flaunt it: they let black and yellow, white, orange, or red colors do the talking. It gives them immunity to assault while they are drinking nectar, sipping water, pausing to bask or groom themselves, hunting prey, or courting. Wasp mimics freely associate with wasps on flowers, at aphid colonies, puddles and pools, and on sunlit vegetation, tree trunks, and logs. Wingless velvet ants scurry over the ground without worry, but beetles, bugs, and even spiders mimic them in appearance and gait.

The clever masquerade can be used to more nefarious ends in some instances. Aggressive mimicry is the "wolf in sheep's clothing" ploy. Insects that normally have nothing to fear from wasps may ignore the mimic and meet an unfortunate end. Robber flies are swift predators that often resemble wasps. Thick-headed flies, which can be easily mistaken for potter wasps, are parasitoids of adult solitary wasps and bees that peacefully co-exist with potter wasps.

Mimic and killer
A robber fly, *Callinicus pictitarsis*, from southern Arizona, U.S.A. looks like a wasp at first glance. It can also kill wasps, as it has here.

Flies: Extraordinary Fakers

Flies of every stripe are eaten by many other animals, including wasps. Flies that are striped *like* wasps have a bit more protection going for them, how do you tell a mimic from the real thing? To start with, flies have only one pair of wings, wasps two. Flies usually have short antennae and enormous compound eyes that take up most of the head. They have sponging mouthparts or an obvious proboscis in contrast to the chewing mouthparts of wasps. Despite such stark differences, flies still manage to fool the world.

Bee Fly
This bee fly, a *Systropus* sp. from Indonesia, has the thin body and long antennae of a wasp, but only one pair of wings.

Flower flies (aka hover flies) in the family Syrphidae are nearly all wasp or bee mimics, appropriate for their out-in-the-open lifestyle of nectar-drinking. These masters of disguise put secret agents to shame. Antennae too short? I'll wave my front legs in front of my face. Eyes too big? I have patterned eyes so you don't know where they end and the rest of my head begins. Yellowjackets fold their wings lengthwise? The leading edge of my wing is dark to mimic that fold. Syrphid flies can have a constricted abdomen to resemble a "wasp waist." They can fly in a weaving pattern like a yellowjacket queen. When seized, some species pretend to sting. Species that mimic solitary wasps may run and flick their wings to reinforce the disguise. The theory that some fly species *sound* like wasps (audio mimicry) has become suspect given recent, rigorous, and high tech analysis. Flower flies are decent pollinators and, in the larva stage, many are voracious predators of aphids.

Female thick-headed flies (Conopidae) accost a solitary wasp or bee in mid-air, then jam an egg between the victim's abdominal segments. The fly maggot that hatches feeds as an internal parasitoid, sucking the host's blood before feeding inside its thorax, eventually killing it. It pupates in the host wasp's abdomen.

Robber flies in the family Asilidae prey on other insects. The typical asilid perches on a leaf, twig tip, log, or stone and scans the sky for a target. It darts out, grabs the hapless insect, pierces it with a knife-like beak, and returns to its perch to feed. Mydas flies look like giant robber flies, but are not predators. One of the world's largest flies, *Gauromydas heros*, mimics tarantula hawk wasps in South America.

Mydas Fly
A large *Diochlistus* sp. mydas fly from Queensland, Australia, is a convincing mimic of a spider wasp (right). These flies are harmless, and feed mostly on flower nectar.

Syrphid Flies
This flower fly, *Spilomyia* sp. (right), waves its front legs to mimic the long antennae of a yellowjacket. More typical syrphids (below), are betrayed by short antennae, enormous eyes, and no jaws.

Some members of the families Stratiomyidae (soldier flies), Bombyliidae (bee flies), Rhagionidae (snipe flies), Therevidae (stiletto flies), Xylomyidae (xylomyids), Xylophagidae (xylophagids), Micropezidae (stilt-legged flies), Pyrgotidae (pyrgotids), Tephritidae (fruit flies), Ulidiidae (picture-winged flies), Platystomatidae (signal flies), Tabanidae (horse flies, deer flies, clegs), Tachinidae (tachinid flies), Bibionidae (March flies, harlequin flies), and Tipulidae (crane flies), are also startling and beautiful wasp mimics.

Beetles: Unlikely Imposters

Few insects are as different to wasps as beetles, yet there is a great diversity of wasp mimics among the Coleoptera. Both insects have chewing mouthparts, but the similarities end there. Beetles, after all, have the first pair of wings hardened into stiff or leathery plates (the elytra). In flight, both pairs of wings are raised, but only the hind pair are responsible for propulsion.

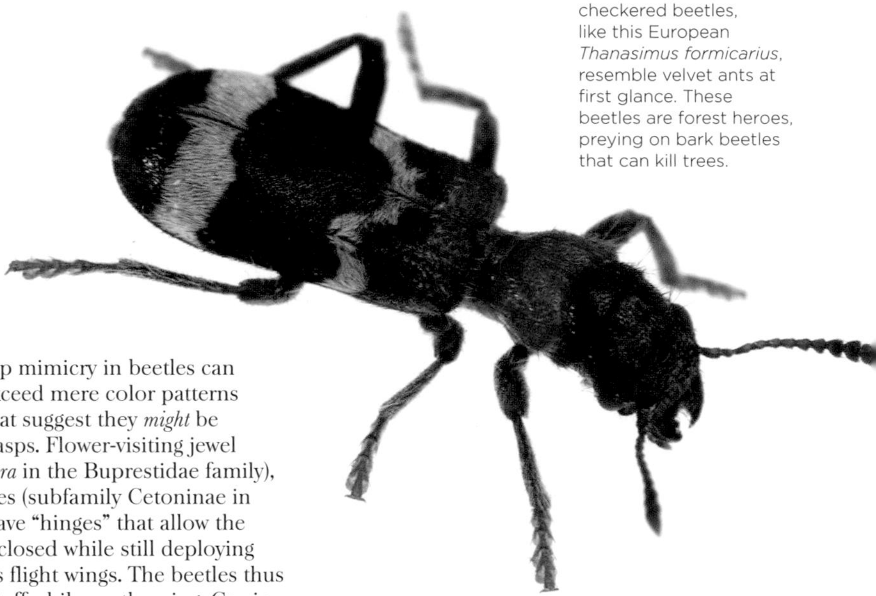

Velvet Ant Mimic
Colorful, fast-running checkered beetles, like this European *Thanasimus formicarius*, resemble velvet ants at first glance. These beetles are forest heroes, preying on bark beetles that can kill trees.

Wasp mimicry in beetles can exceed mere color patterns that suggest they *might* be wasps. Flower-visiting jewel beetles (*Acmaeodera* in the Buprestidae family), and scarab beetles (subfamily Cetoninae in Scarabaeidae), have "hinges" that allow the elytra to remain closed while still deploying the membranous flight wings. The beetles thus maintain their bluff while on the wing. Carrion beetles (Silphidae), reverse their wing covers in flight to enhance a wasp- or bee-like vibe. Rove beetles (Staphylinidae), and some longhorned beetles (Cerambycidae) and soldier beetles (Cantharidae), have short wing covers that leave the flying wings exposed to help carry off the hoax. Rove beetles conceal their flying wings by folding them like an origami project, but in the air larger species are easily mistaken for wasps. *Acyphoderes sexualis* (Cerambycidae) of Panama normally mimics ants, but by flaring its flying wings it transforms into a "wasp" just prior to, and during, flight, dangling its legs while airborne, as do some social wasps.

The subfamily Lepturinae (Cerambycidae) are known as flower longhorned beetles, and many resemble wasps through similar color patterns, long legs, and sometimes elongated ovipositors suggestive of a sting. Other longhorned beetles are also convincing wasp mimics. The hickory borer, *Megacyllene caryae* and locust borer, *M. robiniae*, are two examples from North America. Scientists have discovered that these beetles, when assaulted, release spiroacetals, chemical compounds present in the alarm pheromones of social vespid wasps. Added incentive to drop your prey would be the possibility that it was summoning stinging nestmates to the rescue.

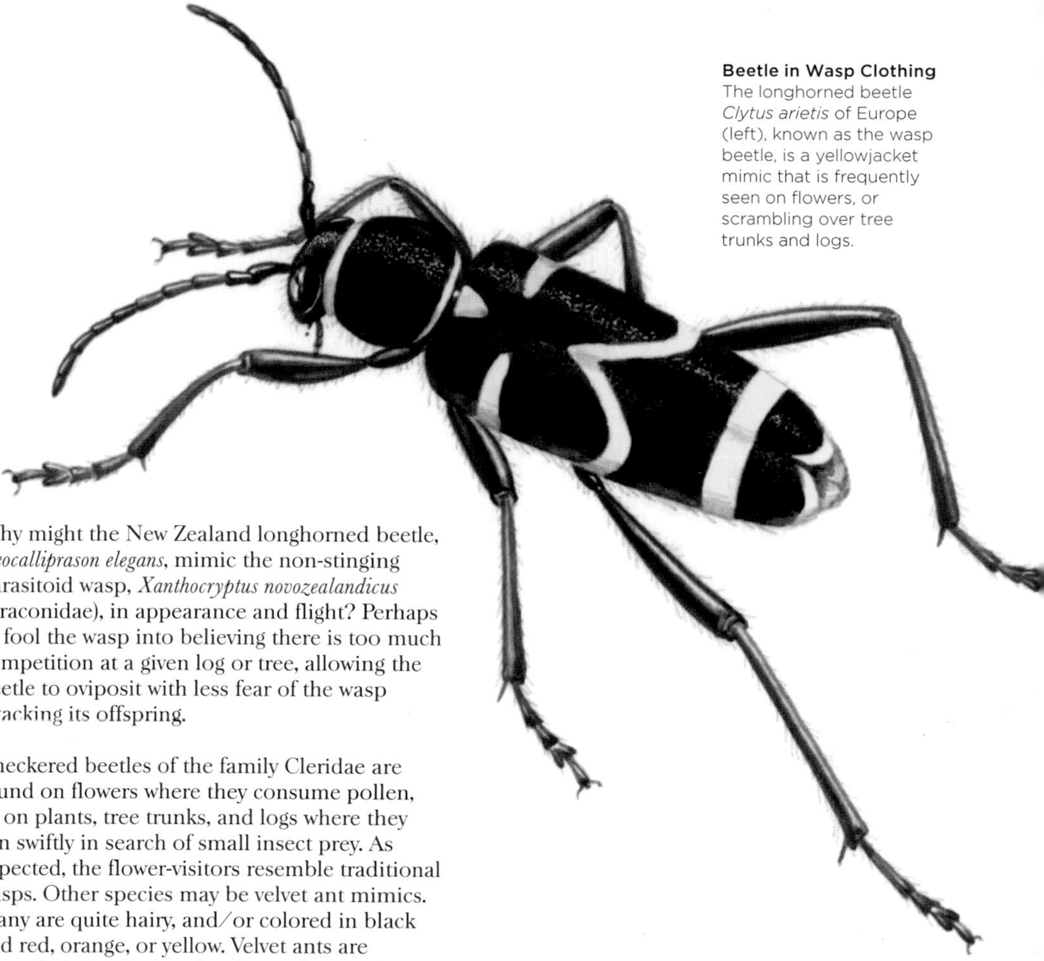

Beetle in Wasp Clothing
The longhorned beetle *Clytus arietis* of Europe (left), known as the wasp beetle, is a yellowjacket mimic that is frequently seen on flowers, or scrambling over tree trunks and logs.

Why might the New Zealand longhorned beetle, *Neocalliprason elegans*, mimic the non-stinging parasitoid wasp, *Xanthocryptus novozealandicus* (Braconidae), in appearance and flight? Perhaps to fool the wasp into believing there is too much competition at a given log or tree, allowing the beetle to oviposit with less fear of the wasp attacking its offspring.

Checkered beetles of the family Cleridae are found on flowers where they consume pollen, or on plants, tree trunks, and logs where they run swiftly in search of small insect prey. As expected, the flower-visitors resemble traditional wasps. Other species may be velvet ant mimics. Many are quite hairy, and/or colored in black and red, orange, or yellow. Velvet ants are also mimicked by tiger beetles and ground beetles (Carabidae).

The Wasp Beetle
The longhorned beetle *Clytus areietis* matches most common wasps in size, ranging from 0.35–0.7 inches (9–18mm) in length. All the better to travel incognito.

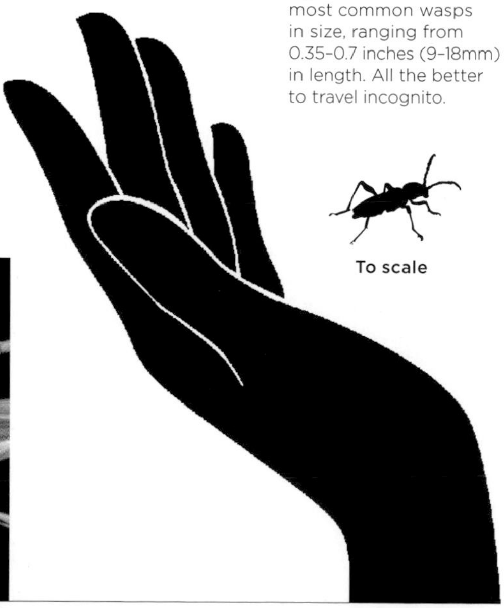

To scale

Flower Longhorned Beetle
Most members of the subfamily Lepturinae (family Cerambycidae), like this *Typocerus zebra* from the U.S.A. (below), are marked in black and yellow. They also frequent flowers like wasps do.

True Bugs, False Wasps

Elaborate modifications of body form and shrewd behavior, allow many true bugs in the order Hemiptera to impersonate wasps. True bugs have piercing-sucking mouthparts, the beak-like rostrum usually concealed beneath the insect along its midline. Cicadas, lanternflies, aphids, leafhoppers, stink bugs, assassin bugs, lace bugs, and water striders are familiar examples of Hemiptera.

Assassin Bug
The predatory bug, Zelus vespiformis (above), of Brazil is a mimic of a braconid wasp. It lives in the understory of the rainforest.

Not Wasp
A surprising wasp mimic is this leafhopper, *Lissocarta vespiformis* (right), of the Peruvian jungle. Its posture and slender body allow it to pass itself off as a social wasp.

The assassin bugs in the family Reduviidae are predatory, and quite capable of defending themselves with a painful bite, yet many opt to dress and act as wasps. The most remarkable example may be *Hiranetis braconiformis* of Central American rainforests. As its name implies, it mimics a braconid wasp in the genus *Monogonogastra*, complete with slender body and banded wings. They walk like the wasp, too, and splay and flick their wings as they move. Should a predator be more curious, the bug does an extraordinary thing: it extends one hind leg beyond the tip of its abdomen to simulate an ovipositor. *Graptocleptes* species in South America are similar mimics.

Zelurus assassin bugs of Central America pass themselves off as tarantula hawks or other spider wasps, bobbing their antennae and/or flicking their wings as they imitate the wasp's stride. *Coilopus* assassin bugs of Central and South America are apparent mimics of social wasps in the genus *Mischocyttarus*. The front lobe of its thorax is a fake wasp head, making up for the pinhead of the assassin bug. The reduviid even has thickened antennae to be more convincing. *Sphodrolestes* species are likewise excellent mimics of neotropical social wasps. Many broad-headed bugs in the family Alydidae mimic ants when they are young, and pretend to be wasps as adults. The transformation of the body from nymph to adult is stunning.

Lissocarta is a genus of South American rainforest leafhopper, but you would never suspect so. The insects look and pose exactly like paper wasps. They compensate for hair-like antennae by extending their front legs in front of their faces. They raise their wings in the posture of a wasp as well. They are not the only leafhopper wasp mimics, as South American *Propetes* sharpshooters look and behave similarly. *Propetes triquetra* was once considered two different species because males and females have two different wasp models that they mimic.

Treehoppers of the family Membracidae are known to mimic thorns, but some tropical species have even more elaborate modifications of the pronotum (top of thorax). The extra lobes on the pronotum of *Heteronotus vespiformis* are utterly convincing of a social wasp, especially with the added feature of clear, flared wings. These South American insects gain protection for both themselves and the offspring they guard by this ruse. Other membracids mimic ants in the same fashion.

No Way
This adult red bug or "cotton stainer" in the family Pyrrhocoridae masquerades as a wasp in the jungles of Borneo. The white antenna bands may mimic its model, or make the antennae look shorter.

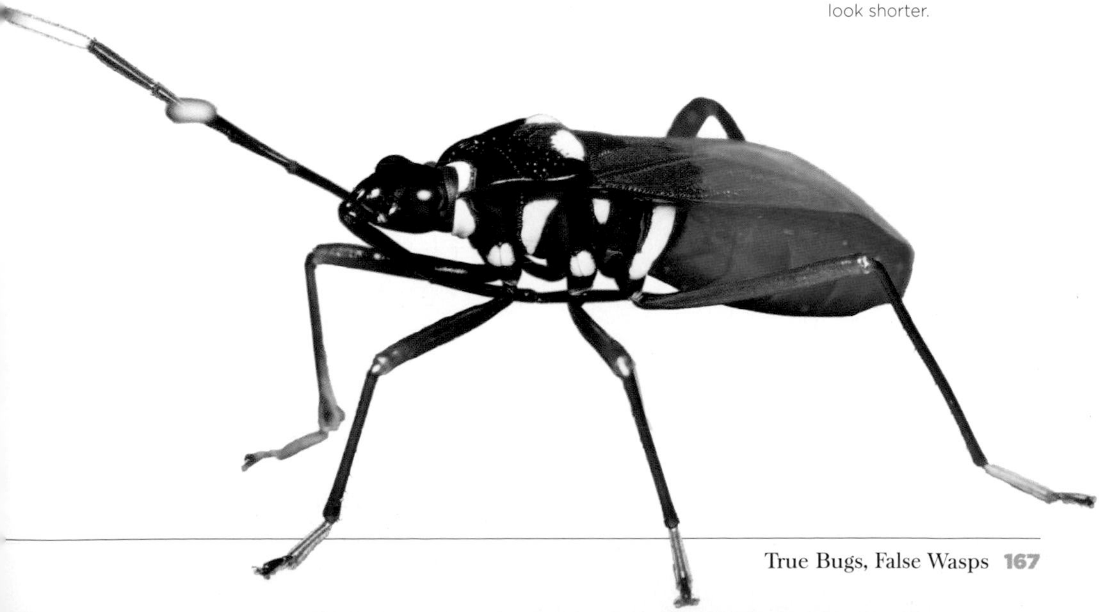

SPOTLIGHT ON

Wasp Mantisfly: Weirdest Wasp Mimic Ever

"Frankenbug" is the word frequently applied to the outrageously unique members of the family Mantispidae in the order Neuroptera. These insects are related to lacewings, antlions, and owlflies, but appear to be the product of a science experiment gone horribly awry—the front half of the body appears to be the shrunken version of a praying mantis. Were that not creepy enough, the life cycle of some species borders on science fiction.

Perfect Mimic
A yellow form of the wasp mantisfly, *Climaciella brunnea*, looks nearly identical to a paper wasp. It even has darkened edges on its wings to mimic the longitudinal folds of its model.

Killer Grip
The mantis-like raptorial forelegs of this mantisfly are built to seize other insects in a lethal embrace. Poised to attack (opposite page), the legs are folded in repose.

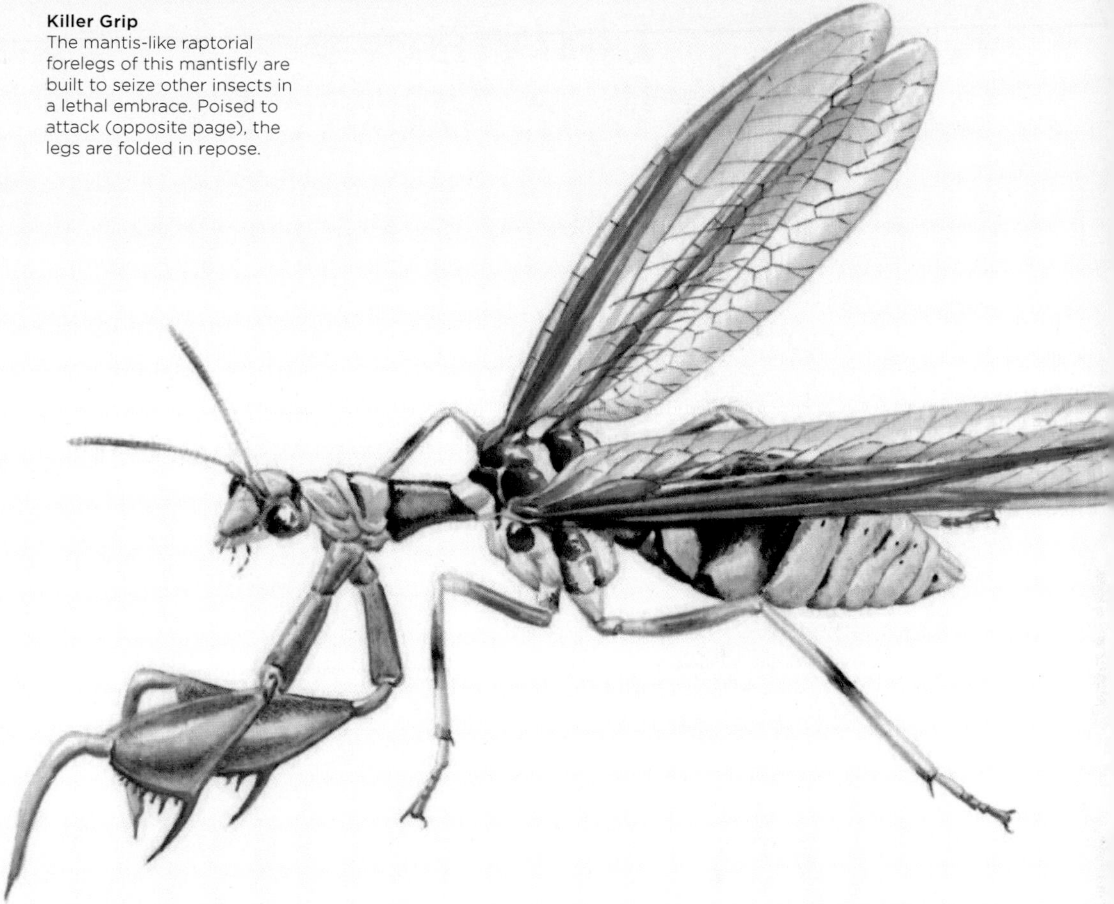

The wasp mantisfly, *Climaciella brunnea*, occurs over much of North and Central America. It is large, at 0.5–0.75 inches (13–19mm) in body length, and is a striking wasp mimic. Even more astonishing is that it is polymorphic in its mimicry. In Costa Rica alone, it mimics four species of paper wasps (*Polistes* spp.), as well as the warrior wasp, *Synoeca surinama.*

Adult wasp mantisflies have two pairs of wings, equal in size, the hind pair concealed beneath the front pair at rest. The leading edge of the wing is darkened to resemble the longitudinal fold of the wings of paper wasps; the version that mimics the warrior wasp has black wings. The raptorial front legs are tucked close to the body so as not to detract from the wasp disguise when the insect is viewed from above. Like mantids, mantispids use those forelegs to seize prey. They often frequent flowers where they pick off pollinating insects and other flower visitors.

The female *Climaciella brunnea* lays scores of eggs, each on a silken stalk. One study found that a female could live for 43 days and lay 3,334 eggs in that time. The larva that hatches descends the stalk and shortly thereafter assumes an erect posture. It is waiting for a passing spider. If it successfully gloms onto a host, the larva's journey is just beginning. Clinging to the rear edge of the spider's carapace, it waits for its host to spin an egg sac. Should the spider be male, the larva must transfer to a female during copulation. The larval mantispid must sneak into an egg sac under construction, as it cannot penetrate a completed one. Once inside it metamorphoses into a sluggish grub that eats the spider eggs. Finished, it spins a cocoon and pupates. An adult mantispid eventually chews its way out.

Katydids: The Sheep in Wolf's Clothing

We usually think of katydids (aka bush crickets) as nocturnal, well-camouflaged, green or brown grasshoppers with long antennae. The males call to females through stridulation, rubbing a "file" on the shoulder of one forewing against a "scraper" on the other wing. Katydids can be the prey of stinging parasitoids like digger wasps in the genus *Sphex*, but a few katydids have evolved to look like wasps themselves.

The genus *Aganacris* is represented by three species of rainforest wasp mimics in South America. There were once thought to be more, lumped in the genus *Scaphura*, but some were described from males or females only. The trouble is that the species are sexually dimorphic: males model themselves on one wasp, while female taking after another. These katydids are more active by day, moving about and flaunting their bright colors. They have also been observed feeding on small insects during daylight hours. At night they are more wary, like normal katydids are, and engage in the courtship and mating behavior expected of strictly nocturnal species.

Utterly Convincing
The colorful katydid *Aganacris insectivora* looks and behaves so much like a tarantula hawk that even seasoned entomologists are hesitant to approach it. This one was photographed in a rainforest in Panama.

Tarantula hawk wasps in the genera *Pepsis* and *Hemipepsis* are the models for adult *Aganacris velutina*. Both the wasps and the katydids are large, and the katydids have short, thick antennae, just like the wasps. They do not bob them constantly, though, and they walk slowly and deliberately like most katydids. Males of *Aganacris nitida* and *Aganacris sphex* probably mimic thread-waisted wasps in the genus *Eremnophila*, complete with a terminal wing spot designed to resemble the wasp's abdomen. The nymphs of *Aganacris* may resemble other wasps entirely, complete with faux wings in the pigment on their backs. An unidentified, colorful specimen from Peru clearly mimics an ichneumon wasp, with white-banded antennae that it vibrates constantly, like the wasp.

Eleven katydid species in the genus *Scaphura* also occur in the neotropics. They have long antennae, as they should, but the first segments are thickened and colored, creating the illusion that they are short. At least one species, *Scaphura nigra*, is polymorphic: one form mimics tarantula hawks in the genus *Pepsis*, while another resembles a spider wasp in the genus *Entypus*. A third form is a *Polistes* paper wasp mimic. Polymorphic mimicry and camouflage is not uncommon in tropical katydids in general. When disturbed, *Scaphura* can assume an intimidating threat posture, with abdomen curled forward and wings flared.

Other potential wasp mimcs include the genus *Quiva*, also found in the New World tropics, and *Veria colorata* of Australia. The latter species may not resemble a wasp in physical appearance, but exhibits wasp-like behavior. The apparent absence of wasp-mimicking katydids in Africa and Asia is rather puzzling.

Great Acting!
This katydid, *Scaphura* sp. of Brazil, begins behaving like a wasp in posture and gait when it feels threatened. The impression is usually good enough to keep it alive.

Moths: Surprising Scammers

How can a fly-by-night insect, covered in scales, look anything like a wasp? Why would it bother? It turns out that plenty of moths in the order Lepidoptera are day-active, and therefore benefit from mimicry. Take a closer look at that flower visitor, or the "wasp" at the edge of the mud puddle. Wasp mimics are found in at least 11 families of moths.

Colorful...Moth?
The common wasp moth, *Eressa angustipenna*, also known as the black-headed wasp moth, is found in Australia and the Philippines. Its colorful wardrobe may indicate it is toxic for predators to eat.

To embellish their wasp costume, many species have clear wings, or at least large patches with a total absence of scales. Not all *wasps* have clear wings, so other mimicking moths settle on metallic blue or blue-black wings. Most moths hide their abdomens beneath their folded wings at rest, but wasp mimics typically flare their wings, exposing abdomens that may be banded like a generic wasp.

"Wasp moths" are a subset of tiger moths (family Erebidae, subfamily Arctiinae), and they include stunningly accurate forgeries of wasps, to the point where the model wasp is easily identified. *Pseudosphex laticincta* impersonates a *Mischocyttarus* and a *Polybia* species, both of which are nearly identical in their own shared color pattern. The moth is betrayed only by its pectinate (comb-like) antennae. Meanwhile, *Myrmecopsis strigosa* is a flawless mimic of *Parachartergus apicalis*, another social wasp of the tropical Americas. Wasp moths cannot sting, but most are toxic due to poisons sequestered from plants they fed on as caterpillars.

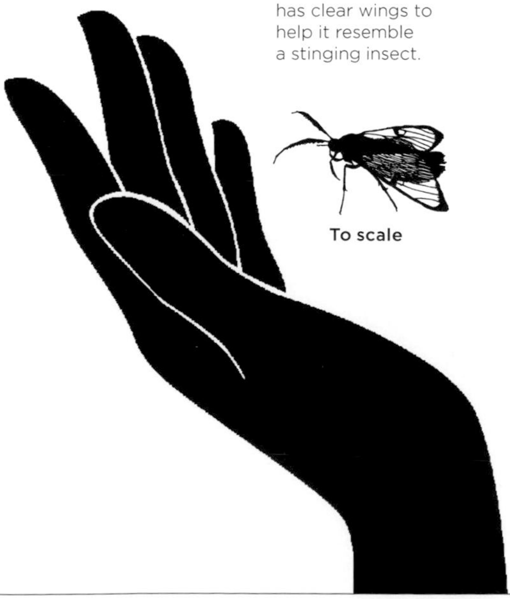

Transparent Wings
This wasp moth from northern Peru has clear wings to help it resemble a stinging insect.

Blue Beauty
The polka-dot wasp moth, *Syntomeida epilais*, is a tropical and subtropical day-flying insect that resembles a real wasp. The caterpillar stage feeds on oleander, so the bold colors warn of its toxicity.

Sphinx moths, also known as hawkmoths, are members of the family Sphingidae. Most are large insects. Smaller species like those in the genus *Hemaris* are fabulous mimics of bees. A few of the bigger species mimic wasps. The nessus sphinx, *Amphion floridensis*, is a common American species that is believed to mimic cicada killer wasps in the genus *Sphecius*. Day-active sphinx moths hover in front of flowers, extending their proboscis into the corolla to pump out nectar. Few wasps are able to hover.

Equally conspicuous, and no less interesting moths include members of the family Zygaenidae. These are the leaf skeletonizer moths of North America, and the burnet moths or forester moths of the Old World. The genus *Harrisina* is familiar in the U.S.A. as a defoliator of grape leaves. Like the wasp moths, many accrue defensive chemicals as larvae that feed on toxic plants.

Additional families (and genera) with wasp mimics include: Glyphipterigidae (*Cotaena*), Stathmopodidae (*Dolophrosynella*), Heliodinidae (*Anypoptus*), Tineidae (*Vespitinea*), Agaristidae (*Cocytia*), and Limacodidae (*Nagoda*, *Nagodopsis*, and *Pseudonagoda*). Male bagworm moths in the genus *Thyridopteryx*, family Psychidae, are excellent bee and wasp mimics, especially in flight. Remarkably, the female bagworm never leaves her silken bag, and becomes sexually mature without otherwise changing much from the caterpillar stage.

PROFILE

Clearwing Moths

Family Sesiidae

The entire moth family Sesiidae is comprised of dedicated wasp mimics. Globally, there are 160 genera and 1,452 known species. New species are discovered regularly, even in North America and other well-studied regions. The larvae are unusual for caterpillars in that they are borers, tunneling in stems, roots, or wood of the host plant. At least one Brazilian species, *Neosphecia cecidogena*, forms galls on a vine (*Cayponia pilosa*). We know the most about species of economic importance, like the squash vine borer, *Melitta curcurbitae*, of North America. The life cycle of many species requires two years.

Fake Hornet
The grape root borer moth, *Vitacea polistiformis* of North America, is a false wasp, but a very real pest, on occasion, of grapes. The caterpillar stage bores in the roots.

When ready to liberate itself to adulthood, the pupa works its way to the surface of the host, erupting through the stalk, shoot, or trunk. When the moth emerges, its wings are completely scaled, like any other moth. During its maiden flight, most wing scales are dislodged and drift through the air, leaving the moth with the clear wings that give the family its common name. The empty pupal cases left behind are often more conspicuous than the moths, as they remain firmly lodged and take time to wither and disintegrate.

Sesiids are mostly diurnal, or crepuscular (active at dusk and dawn), but a few species are attracted to lights at night. At rest, they are not necessarily the most convincing wasp mimics, though the thickened antennae and body posture may be enough to encourage a predator to move on to safer prey. A few species have elongated scales on the rear of the abdomen that bear a crude resemblance to an ovipositor. It is when they are airborne that these moths elevate their game to new heights. They are nearly impossible to discern from the specific wasps they mimic when in flight, duplicating the nuances of their models without peer.

Family Sesiidae

SPECIES	~1,370
DISTRIBUTION	Worldwide except Antarctica, mostly tropical
SIZE	~0.3–1.9 inches (~8–48mm)
AMAZING FACT	Most species fly during the day

Models for clearwing moths run the gamut from ichneumon and braconid wasps to yellowjackets, hornets, and paper wasps, as well as spider wasps. The moths are scarce enough that their model wasps almost always outnumber them, increasing the effectiveness of their mimicry. Scientists studying these moths often deploy synthetic pheromones to mimic the alluring scent of female moths, often drawing surprisingly large numbers of males. Accidentally spilling pheromone on yourself is a great way to deter the opposite human sex, however.

Wasp Costumes
A female (left) and male (above) yellow-legged clearwing, *Synanthedon vespiformis*, of central Europe and the Mediterranean. They fly by day and sip flower nectar. The caterpillars bore in woody plants and can be orchard pests.

Actual size

Hornet Moth
Sesia apiformis is native to Europe and the Middle East, but has become established in North America. The caterpillar stage bores in the roots, bark, and lower trunk of poplar trees.

SPOTLIGHT ON

Spiders: No Wings, No Problem

Eight-legged arachnids might be the last organisms you could imagine being wasp mimics, but remember that velvet ants are wingless wasps. Many spiders that wander freely and do not spin webs have adopted ant mimicry, so it is not an evolutionary stretch to include velvet ants; a surprising number of ground-dwelling and arboreal (tree-dwelling) spiders take after velvet ants.

Jumping Spider
This *Phiale formosa* from Costa Rica, family Salticidae, is very likely a mimic of a velvet ant.

Coming or Going?
This jumping spider, *Orsima ichneumon*, is headed west, but it wants you to believe it is a velvet ant headed east. The spinnerets are fake antennae, the abdomen is constricted into "head" and "thorax."

Spider families with certifiable mutillid mimics include: jumping spiders (Salticidae), ant mimic spiders (Corinnidae), and ground spiders (Gnaphosidae). Striking color combinations of black, red, orange, yellow, and white are common in these mimics. German entomologist and arachnologist Ferdinand Karsch may have been the first to recognize this unique mimicry in publishing his description of the female *Coenoptychus pulcher* (Corinnidae) of Ceylon (present day Sri Lanka) in 1891. Revisiting this species has revealed that the female spider mimics a *Bischoffitilla* species of velvet ant, while the male spider is a mimic of *Trogaspidia villosa*.

Jumping spiders may represent the pinnacle of wasp mimicry. Females of *Phiale guttata* in Panama have two color forms. One is red and black, the other yellow and black, each presumably mimicking a different velvet ant. Males of *Cosmophasis nigrocyanea* are ant mimics, like many salticid spiders, but the females are mutillid wasp mimics. *Phidippus apacheanus*, a large, hairy, orange and black jumping spider of the U.S.A. is undoubtedly a mimic of *Dasymutilla* velvet ants. Other *Phidippus* species may be mutillid wannabes, too, but often it is only the male that has bright coloration in adulthood, females and juveniles being cryptic by comparison.

Two examples of wasp mimicry by jumping spiders defy belief. *Depreissia decipiens* of Borneo lives high in rainforest trees. The pedicel connecting the cephalothorax and abdomen is greatly elongated, evoking the image of a social wasp like a *Belonogaster* species. *Orsima formica* of Malaysia is a velvet ant mimic in reverse. Its bi-lobed abdomen is the head *and* thorax of the "wasp," complete with fake antennae in the form of elongated spinnerets.

Despite its common name of "wasp spider," the large and colorful *Argiope bruennichi* from Europe, northern Africa, and parts of Asia, is not a wasp mimic. The black- and yellow-banded abdomen is instead a lure that attracts insect prey.

Model Wasp
Female velvet ants, like this *Dasymutilla occidentalis*, have a wicked sting, so are mimicked by many harmless arthropods, from ground beetles to ground spiders.

9

Enemies of Wasps
Subverting the Sting

Wasps Are Not Invincible

Despite an armored exoskeleton, and often a venomous sting, wasps are not impregnable. They are vulnerable to a host of enemies at every stage of life. Obsessed as we are with our own human mortality, we fail to remember that other animals face the same hazards: predators, parasites, and pathogens. Wasps can succumb to abiotic factors, too. Excessive heat or cold can prove fatal, while a glitch in the process of molting can mortally cripple them.

Vertebrate predators of wasps abound in most ecosystems. Birds are skillful predators of nearly all insects, and dodging a stinger presents little challenge. A few species are even brave enough to raid entire nests of social wasps. Mammals, too, will consume colonies wholesale, while lizards are not above making a meal of adult or immature wasps either. Human beings, of course are the worst of all. We destroy countless nests of solitary and social wasps in the course of pest control, gardening, development of new industrial parks, businesses, and residential subdivisions. We murder individual wasps unintentionally simply by driving our vehicles.

However, the greatest threat to a wasp's life is usually another invertebrate. An extraordinary number of other insects, plus spiders, mites, nematode worms, and other spineless animals seem to be waiting in line to attack. A wasp's biggest fear may be…another wasp. Parasitoids have parasitoids, often several. Parasitoid wasps that hunt other dangerous animals, like spiders, risk becoming a victim themselves at the jaws of their quarry. The fearlessness of wasps in the face of such potentially tragic outcomes is one more thing to admire about them.

Some enemies are more insidious and persistent than others. Satellite flies are a constant threat to the offspring of most solitary wasps, stalking the mother wasp while she creates her nest, and then stealing any opportunity to deposit their own maggot offspring inside. *Melittobia* wasps are abundant, tiny parasitoids that overwhelm the helpless immature stages of much larger wasps. Few parasites are as twisted as the strepsipterans, so enigmatic that scientists are not even certain how to classify them. The bizarre insects bulge from, and contort, the abdomens of their hosts.

The idea that a wasp can contract a disease, get sick, and die from the ordeal is a foreign concept to most of us. Yet pathogens in the form of fungi, microbes, and viruses are common enemies of all insects, and sometimes used by us to control pest species.

There are no "waspitals," and no first responders, so vigilance and avoidance are a wasp's major weapons against assault by its enemies. Amazingly, wasps soldier on, and even thrive, constantly adapting their behavior and evolving new survival strategies.

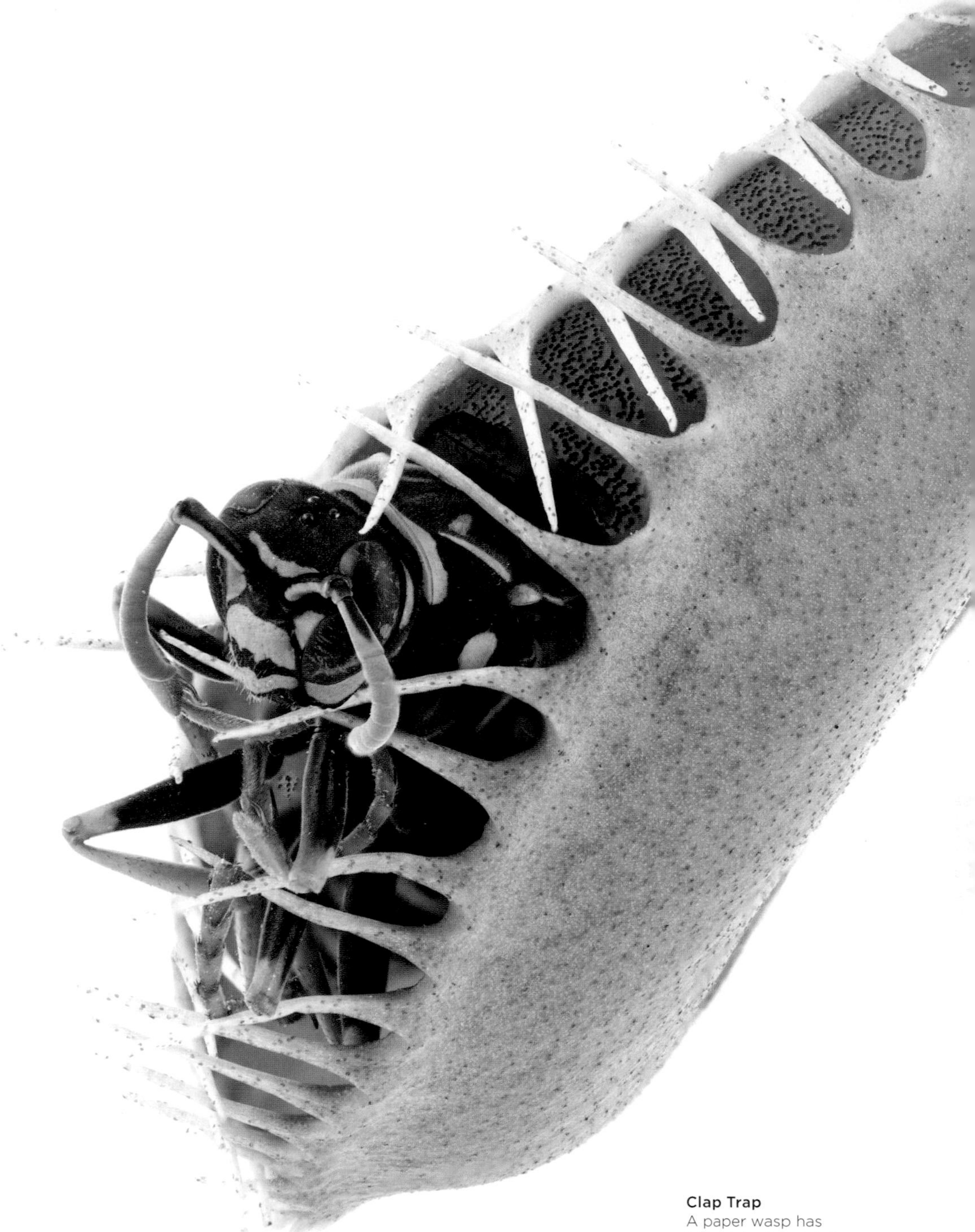

Clap Trap
A paper wasp has fallen into the clutches of a venus flytrap. The carnivorous plant grows in nitrogen- and phosphorous-poor soil in bogs, and so must augment its nutrition requirements in a macabre fashion.

Vertebrate Predators

The brood (eggs, larvae, and pupae) of wasps are coveted morsels of protein for a variety of birds, mammals, and sometimes lizards. Even adult wasps will be snatched, but colonies of social wasps represent a windfall of protein to any animal willing to endure the stings of the adult insects. Wasps have other defenses, too.

Wasp Predator
The red-bearded bee eater of southeast Asia is an expert at snatching and dispatching wasps in its jungle habitat.

Wasps are not known for evasive flight maneuvers, and often fail to detect avian predators in time to avoid them. Flycatchers, tanagers, shrikes, ant-shrikes, magpies, crows, starlings, orioles, catbirds, bluebirds, sparrows, warblers, chickadees, great tits, and aptly-name bee-eaters, are among the birds documented to prey on wasps. The beak of a bird is impervious to a sting, but birds like tanagers forcibly tear out the sting before consuming the insect. Woodpeckers may break into galls to extract the larval wasps within.

The summer tanager, gray-headed kite, red-throated caracara, and the honey buzzard, dare to attack entire nests of social wasps. The honey buzzard claws its way into mostly underground nests of yellowjackets. It has dense facial feathers that help deflect stings, and also secretes a kind of chemical tranquilizer that disorientates the wasps as they attempt to defend the colony. The caracara makes aggressive, repeated attacks until the wasps concede defeat and abandon the nest. Over 75 percent of the bird's diet can be wasp brood; the rest is mostly millipedes.

Many mammals also attack wasp nests. Bears, badgers, skunks, and sometimes raccoons are particularly talented at digging out and tearing open yellowjacket nests in temperate North America and Europe. In the New World Tropics, bats have been observed consuming the brood of paper wasps in Brazil, while white-faced monkeys raid nests of *Polybia* social wasps in Panama. Rodents gnaw into galls, and raid wasp nests, too.

Geckos are the primary reptilian predators of wasps, especially in tropical and subtropical regions. Skinks, spiny and fence lizards, and occasionally horned lizards are among other known wasp consumers, attacking both adult wasps and the brood of social wasps. Bullfrogs and other amphibians may eat individual adult wasps.

Social wasps attempt to thwart vertebrate attacks by concealing their nests in cavities, constructing them out of reach, or attaching them to flimsy substrates that will not support the weight of a heavy predator. Threatened, the wasps may head-butt an approaching antagonist as an explicit warning that the worst is yet to come. A few species also spray venom from their sting, directing the jet toward the eyes of the adversary. This avoids a potentially suicidal colony defense mission should the sting become lodged in the hide of the enemy, or said foe simply crushes the wasp.

Brazen Attack
A red-throated caracara tears into a nest of social wasps in Panama (above). The promise of a bounty of juicy wasp larvae and pupae is a great incentive for vertebrate predators to brave stings.

Stinging Snack
A common wall lizard, *Podarcis muralis*, has grabbed a yellowjacket in a town in Italy (right).

Invertebrate Predators

Ants, and mantids, and spiders, oh my! Wasps constantly run a gauntlet of predators in the form of other invertebrates. Ironically, their fellow armored arthropods are less concerned with the potential for stings than vertebrate predators. Nests are especially vulnerable, but wasps have ways of coping.

Aerial Assassin
A delta-spotted spiketail dragonfly, *Cordulegaster diastatops*, has plucked a sawfly (Tenthredinidae) off a leaf, and has now come to rest and dine on it along a wooded stream in Canada.

New World army ants (*Eciton* and related genera) are notorious for emptying the nests of social wasps, carting off eggs, larvae, and pupae to feed their own ant larvae. Faced with the overwhelming onslaught, the wasps abandon the nest and establish a new colony elsewhere. Most ant species will attempt to raid a wasp nest when the opportunity presents itself. Paper wasps that found new colonies singly (genera *Mischocyttarus*, *Polistes*, and *Ropalidia*) counter by frequently applying a chemical from abdominal glands to the pedicel from which the nest is suspended. The material apparently repels ants by interfering with the chemical trails they lay down while foraging.

Visiting a flower, or pausing on foliage to groom, can seal a wasp's doom. Camouflaged mantids, assassin bugs, and spiders wait motionless to ambush any careless insect that comes close enough to seize. Mantids are strong enough to simply grab prey and commence feeding. Assassin bugs and ambush bugs (family Reduviidae) are quick enough to grab a victim and inject venom through their rostrum. The prey is immediately paralyzed. Crab spiders (Thomisidae) operate in a similar manner, ending the struggles of prey almost immediately by delivering venom through their fangs.

Web-building spiders may intercept flying wasps in their silken snares, but often opt to cut them free rather than battle fang to sting. Wasps that hunt spiders, like mud daubers, risk having the tables turn if they become entangled in a web while trying to tease out the spider.

There are enemies on the wing, too. Dragonflies may swoop in and grab them. Robber flies, aka assassin flies (Asilidae) often tackle victims in mid-air, using a knife-like proboscis to inject paralyzing venom. The fly then returns to its

perch to feed leisurely. Without the chewing mouthparts of ants and mantids, these true bugs, flies, and spiders must also inject enzymes to begin extraoral digestion, liquifying the internal tissues of the victim, which are then drawn out through the mouthparts.

A wasp's worst enemy may be another wasp. Hornets, and the bald-faced hornet (a type of large yellowjacket in North America), are known to prey regularly on smaller yellowjackets. Small solitary wasps are sometimes the prey of beewolf wasps in the genus *Philanthus*. The female beewolf digs a multicellular burrow underground and provisions each cell with paralyzed bees, normally, but the odd wasp suffices, too.

Beak of Death
An assassin bug in Brazil (above left) has killed a wasp by using its beak-like rostrum to inject paralytic venom and digestive enzymes into its victim.

Red Robber
A robber fly, *Saropogon* sp. (above right), has killed a small wasp in southern Arizona, U.S.A. Many robber flies also look like wasps.

Morbid Mantis
A mantis finishes consuming a yellowjacket that it seized with its lightning-fast front legs (below). The serrated legs hold victims in an inescapable grip.

SPOTLIGHT ON

Stylopids

Twisted-wing parasites are so strange that they have been alternately classified as beetles, or worthy of their own order. Today, they are the order Strepsiptera, with nine families and about 600 species. All are internal parasites of insects. The genera *Xenos*, *Pseudoxenos*, *Paraxenos*, *Paragiaxenos*, and *Eupathocera* in the family Stylopidae attack wasps.

Bulging Parasites
A female northern paper wasp, *Polistes fuscatus*, is afflicted by several stylopids protruding from beneath her abdominal segments.

Miniscule Male
A tiny male stylopid has elaborate antennae, full of sensors tuned to the sex pheromones released by females. A male *Xenos vesparum* appears fragile, but is a capable flier that zeros in on females still lodged in host wasps.

The common name refers to the males—tiny insects with the first pair of wings reduced to leathery stubs that appear "twisted," especially in preserved specimens. The second pair of wings are large and broad. The male has large eyes, and branched or otherwise modified antennae. He lives a few hours, at most a day, and does not feed.

Males use their antennae to receive the pheromone emitted by the female. She bears no resemblance to him. Tucked between abdominal segments of her host wasp, she is wingless, with no legs, and rudimentary eyes. The hardened head and front portion of the thorax (prothorax) protrude from the host, but the rest of her soft body may fill most of the abdomen of the host.

The male uses his genitalia to pierce the cuticle of her brood canal, a trough between her head and prothorax. This is termed hypodermic insemination. The eggs hatch within the female's body, and larvae pour out by the hundreds through the brood canal as active planidia. The planidia typically exit when the host lands on a flower or other substrate likely to be visited by another wasp. Climbing aboard a new female wasp, they are ferried to her nest.

Once inside a nest, one or more planidia bore into the wasp egg or larva and begin feeding on blood (hemolymph). The stylopid larvae grow slowly, in sync with the host such that both enter the pupa stage at about the same time. The puparium of the stylopid forces its way between abdominal segments during the last days of the host pupa stage, or shortly after the adult wasp emerges.

Stylopids rarely kill their hosts, but do exert weird effects. Female wasps that are "stylopized" rarely make nests or hunt prey; and they can have anatomical features of males. Stylopized male wasps may have female features. Both sexes may suffer from smaller body size and/or distorted body parts.

Invertebrate Parasites and Parasitoids

There are more parasites and parasitoids of wasps than there are predators. From mites to beetles, flies, and other wasps, there is no shortage of villains in the wasp storyline.

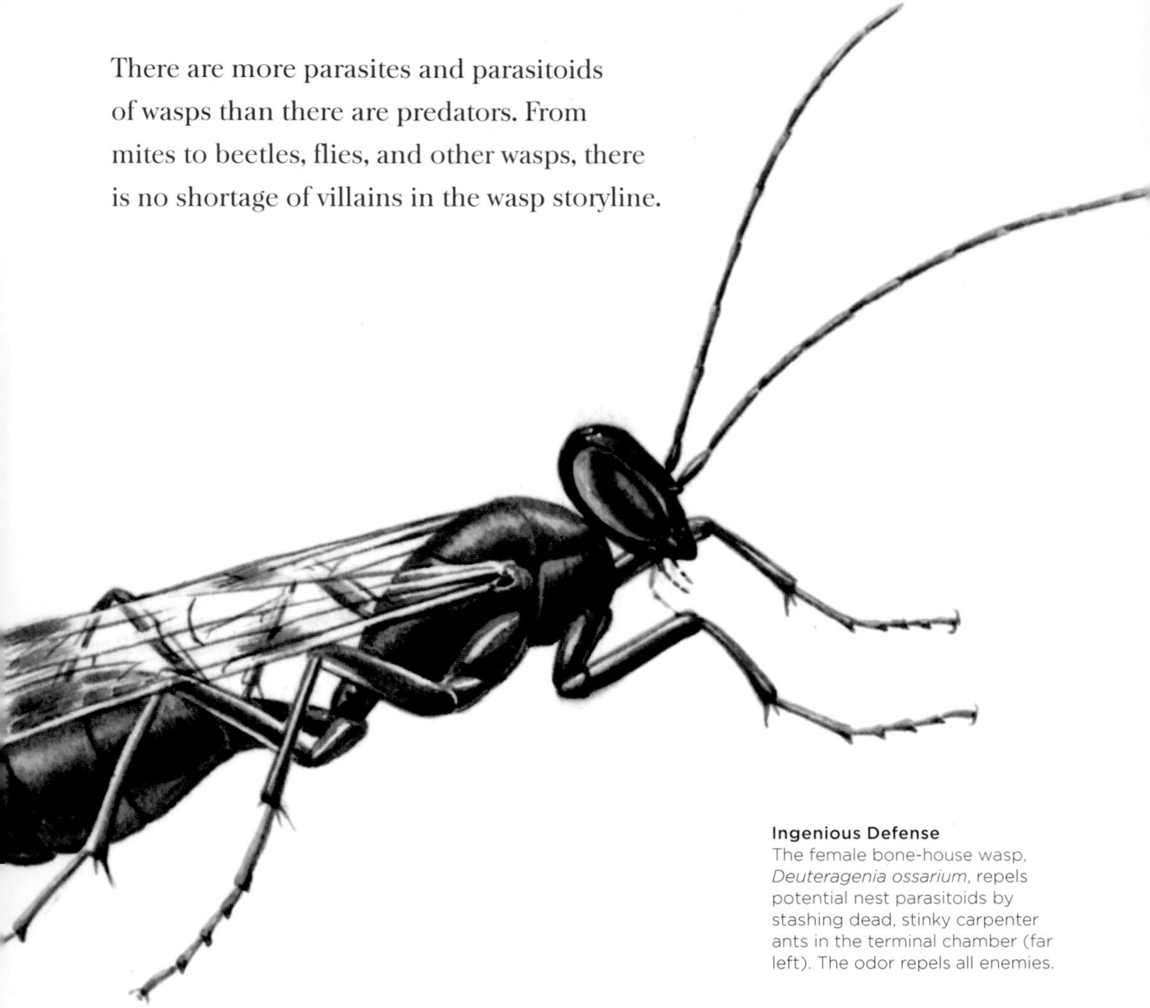

Ingenious Defense
The female bone-house wasp, *Deuteragenia ossarium*, repels potential nest parasitoids by stashing dead, stinky carpenter ants in the terminal chamber (far left). The odor repels all enemies.

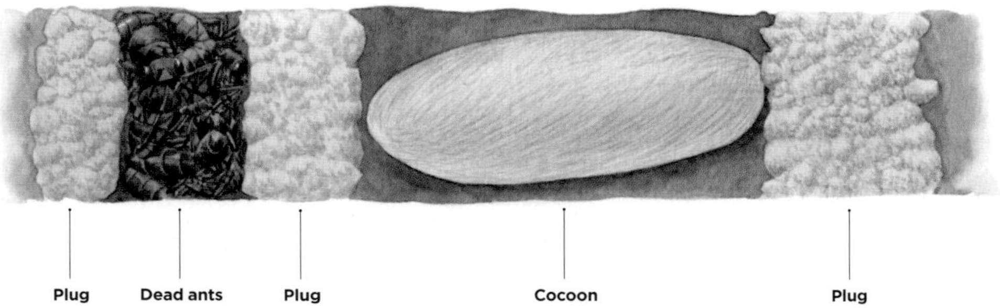

Tiny Hitcher
The first instar larva of a ripiphorid beetle clings to the mouthparts of a spider wasp as it rides back to the wasp's nest. Once there, it starts feeding on its host's offspring.

Mites may be scavengers, kleptoparasites, or parasites in wasp nests. Some, like *Pyemotes ventricosus*, are generalists that associate with almost any insect. Inside wasp nests they feed on the wasp egg, larva, or pupa, or consume prey items intended for the baby wasps. At the other extreme are mites tied to specific hosts. Adult *Ensliniella* and *Vespacarus* mites suck the hemolymph of the larvae of mason wasps, *Allodynerus* and *Parancistrocerus* respectively, but do not kill them. *Kennethiella* mites are killed by female larvae of their mason wasp host, but mites associated with male larvae survive, transferred to adult females through venereal transmission during wasp sex. Many adult mason wasps have built-in "carports" for mites to park in, even those that offer no obvious benefit to the wasps.

Carpet beetles, family Dermestidae, are usually scavengers in wasp nests, but some actively prey on wasp eggs, larvae, and pupae, especially in wasps that nest in linear cavities. Female wedge-shaped beetles (Ripiphoridae) deposit their eggs on flowers and buds. The active larvae that emerge latch onto a female wasp that visits the flower, riding to her nest and infiltrating the cells. Each larva feeds initially as an endoparasitoid, then emerges to feed externally before completing its own life cycle.

Many bee flies are parasitoids or Kleptoparasitoids of solitary wasps. Female bee flies typically hover in front of a wasp nest in a cavity, or over the entrance to a burrow, and fire their eggs into it with forceful thrusts of the tip of the abdomen. The larvae that hatch must seek the host offspring, then wait patiently for them to enter the pre-pupa stage before feeding on them. Scuttle flies, family Phoridae, sneak into nests and lay their eggs. Their larvae feed on prey stored for the host's offspring, occasionally eating the host babies, too.

Entire families and genera of wasps make their living as parasitoids and kleptoparasitoids of other wasps. Spend time watching cavity nesting wasps or mud daubers and you will see a parade of parasitoids, possibly including cuckoo wasps (Chrysididae), velvet ants (Mutillidae), wild carrot wasps (Gasteruptiidae), leucospids (Leucospidae), ichneumons (Ichneumonidae), and numerous micro-Hymenoptera.

Would you believe moths as wasp parasitoids? The female sooty-winged chalcoela, *Chalcoela iphitalis* (family Crambidae) lays her eggs near a paper wasp nest. The hatchling caterpillars crawl into the cells of the comb and feed on wasp pupae or pre-pupae. The moth completes its own metamorphosis by spinning a cocoon inside the cell.

Viruses, Fungi, and Other Pathogens

We are not the only organisms that suffer from infection by microbes, viruses, and fungi. Wasps are not immune, either. How pathogens affect their wasp hosts is still mostly a mystery.

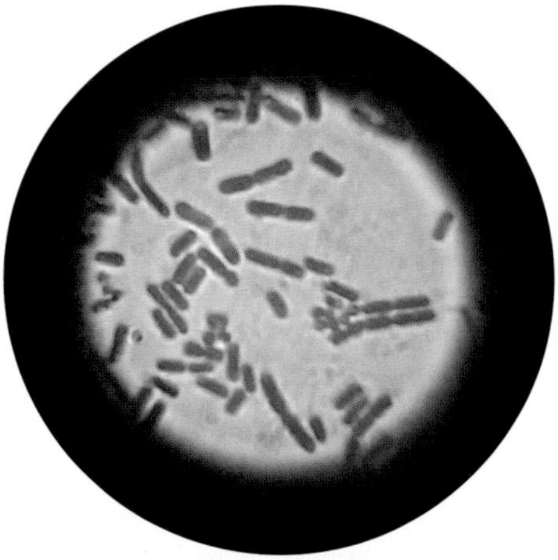

Viruses are cell parasites with a DNA or RNA core and a protein coat called a capsid. Viruses hijack the host cell such that the cell replicates the virus, then bursts to liberate it. Nuclear polyhedrosis viruses (NPVs) replicate in cell nuclei, causing acute fatal diseases. Sawflies in particular are susceptible to rapid population declines when infected.

Deformed Wing Virus (DWV) is probably a complex of closely-related viruses. It can manifest as malformation of wings in adult insects, but asymptomatic carriers appear normal. DWV has been found in seven social and one solitary species of Vespidae. Yellowjackets may contract DWV directly or indirectly when they prey on infected honey bees.

In 2017, scientists published the discovery of a virus in the order Mononegavirales inside the parasitoid *Pteromalus puparum.* The virus increases the lifespan of the wasps, but reduces the number of female offspring produced. Both effects favor the virus, naturally. Increasingly, scientists are finding viruses specific to parasitoid wasps that help the wasps manipulate their hosts (page 95).

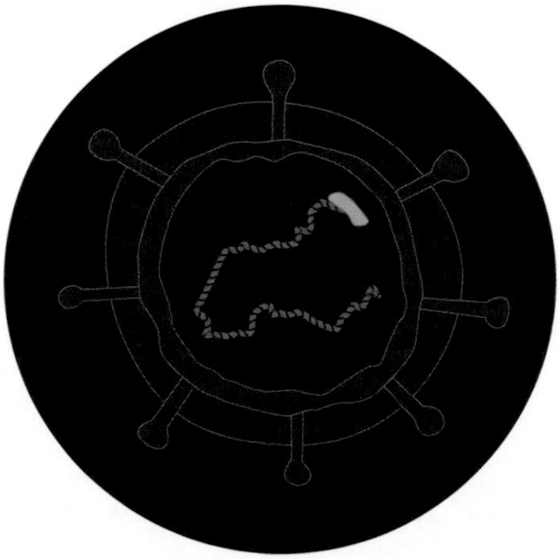

Bacillus spp. and *Serratia* spp. are bacteria pathogenic in insects, but their occurrence in wasps could be from infected prey provided to wasp offspring. *Wolbachia* bacteria are found in roughly 70 percent of insects. Their major effect is manipulation of host sex determination. The spread of *Wolbachia* is female-driven.

Sprouting Fungus
A tropical social wasp (Vespidae) (above) has been killed by an entomopathogenic fungus. The fruiting bodies of the fungus now erupt through joints in the exoskeleton of the host, raining spores to infect other insects.

Going Viral
Bacteria (left, top) and viruses (left) both overwhelm their hosts through rapid reproduction. Viruses hijack host cells and force them to reproduce viral DNA or RNA instead of performing normal functions.

Infected female insects always infect their offspring. In some insects, uninfected females mating with infected males results in death of the eggs. In other cases, infected females, whether mated with infected or uninfected males, results in offspring that are female only. *Wolbachia* facilitate parthenogenesis in many parasitoid wasps by doubling the number of chromosomes, allowing the spawning of female offspring without fertilization.

Pathogenic fungi associated with wasps include *Aspergillus flavus* and *Beauveria* spp., especially in social wasps. *Ophiocordyceps* spp. fungi are the stuff of science fiction. Wasps likely encounter spores while flying, or from water droplets. The best known *Ophiocordyceps* create "zombie ants." Assuming the same mechanism of "mind control" applies to wasps, the fungus produces a cocktail of chemicals that instruct the brain of the host to dictate climbing or flying behavior, putting the host insect about 36 inches (1000mm) above the forest floor, locked in a death-grip to a twig or leaf with its jaws and/or legs. There, high humidity and high temperatures allow the fungus to finish growing. Mycelia and fruiting bodies begin sprouting from every orifice of the host, including leg joints and from between body plates.

PROFILE

Satellite Flies

Sarcophagidae: Miltogramminae

Wasps have "stalkers" in the form of certain flies that follow them to their nests and execute a home invasion. These insect felons belong to the subfamily Miltogramminae in the family Sarcophagidae. Each has its own *modus operandi* for penetrating the defenses of solitary wasps.

Family Sarcophagidae, Subfamily Miltogramminae

SPECIES	~600
DISTRIBUTION	Worldwide except for Antarctica
SIZE	0.1–0.19 inches (3–5mm)
AMAZING FACT	Females lay live larvae instead of eggs

Actual size

Persistent
Satellite flies make up in persistence what they lack in size, repeatedly advancing on a wasp nest until they can successfully deposit their larvae.

Miltogrammines are indiscriminate, opportunistic offenders. *Senotainia trilineata*, for example, has been documented from at least 44 species of ground-nesting wasps, in four families. *Amobia floridensis* specializes in tracking wasps to their mud nests or pre-existing, linear cavity nests. It is known from at least seven crabronid wasp species and five mason wasps.

Instead of laying an egg, the female larviposits. That is, she deposits a live maggot that quickly enters the host nest. Different flies have different approaches. Many are "hole-searchers" or "hole-watchers." These flies have small eyes, enlarged antennae, and probably orient to hosts and their nests visually, and by scent. They take note of disturbed sand, open burrows, and wasp nesting activity. *Metopia luggeri* locates an active nest, perches on nearby vegetation, and waits for a chance to larviposit.

Flies in the genus *Phrosinella* are experts at breaking and entering, digging through closed burrows. True satellite flies are in the genus *Senotainia*. Their eyes have enlarged facets in front and center, the better to pursue their hosts visually. These daredevils follow a prey-laden female wasp into the entrance of her nest, and deposit a larva on the prey while she is carrying it. *Hilarella*, *Spenometopa*, and *Taxigramma* are "stalkers" and "lurkers." They watch for the wasp to enter with prey, or begin nest closure, then dart in to larviposit at the threshold of the burrow.

Once inside the host nest, the maggot turns its attention to food. It may kill the host egg or first instar larva outright, or simply begin feasting on the prey provisioned by the mother wasp, thereby starving the wasp's offspring. Once it has completed its growth, the maggot creates a capsule-like puparium from its last larval exoskeleton and pupates inside. The adult fly pops a weakened suture at the top of the capsule, crawls out and begins the life cycle anew.

Wasps are not without home security to outwit satellite flies. Many ground-nesting wasps create "accessory burrows" that lead nowhere, for example.

Waiting Game
Satellite flies are an ever-present threat to the nests of most solitary wasps (left). Watch any wasp digging her nest and you will see several of these tiny parasitoids darting about.

Surveillance
A satellite fly on a leaf is watching the activity of a cavity nesting solitary wasp (right). Grooming her front legs makes her look like a devious movie villain rubbing his hands together.

PROFILE

Little Nest Pests

Genus *Melittobia*

If wasps had the equivalent of household cockroaches, it might be the parasitoid wasps in the genus *Melittobia*, family Eulophidae. Killer cockroaches at that. You can't leave the kids in the nursery with these blood-suckers around. There are 12 species, but life histories are known for three. They are notoriously inbred, have overlapping generations, exhibit sexual dimorphism—with different female body types according to generation—and have strange behaviors.

Bizarre Polymorphism
Illustrated here is a "normal" male (facing page, top left), and a "normal" female (top right), as well as a blind stubby winged male (below left), and a short-winged female (below right). The shorter-winged individuals represent the second generation produced by the "normal" generation.

Wasps and bees are the principal hosts for *Melittobia*, but these miniscule wasps, less than 0.08 inches (2mm), are adaptable enough to exploit almost any insect host. This makes them pests of insect cultures in laboratories, despite the fact that the female wasps mostly walk or hop, rarely flying. Males have wing stubs.

The female *Melittobia* is able to tunnel through mud nests, nest plugs, and cell partitions to reach the mature host larva she seeks. She feeds by repeatedly stabbing the larva with her ovipositor and sipping the leaking hemolymph for several days before depositing her eggs. The first dozen or so larvae that hatch develop rapidly as external parasites of the host. They complete development in two to three weeks, but are not like their parents. This fast-growing generation is composed of less than three blind males with shorter wing stubs than normal, and numerous larger-bodied females with short wings. Those siblings mate with each other. Mom and her daughters now lay eggs together. The offspring of both combine to completely devour the host, and yield a "normal" generation of long-winged females and normal males, in about 90 days.

Despite the wealth of females, males often fight, even to the death. Males also initiate courtship, producing a pheromone attractant and engaging in ritualized leg raising and lowering, wing fluttering, and stroking of the female with legs and antennae. Females indicate their consent to sex only at the conclusion of the performance.

How do *Melittobia* leave the host nest? Chewing. In at least some species, a female will sting the wall of the nest. The venom serves as a chemical recruitment for other females to join her at that one location and together they eventually breach the mud wall or plug. Males may crawl to an adjacent host nest, or wait for new females to arrive.

Genus *Melittobia*

SPECIES	13
DISTRIBUTION	Worldwide except for Antarctica
SIZE	0.07 inches (<2mm)
AMAZING FACT	Inbreeding is rampant in these wasps

Actual size

"Normal" is Relative

A "normal" female of *Melittobia* has long, functional wings and is the dispersing form of this tiny parasitoid in the family Eulophidae. Most mud daubers and cavity nesting solitary wasps are vulnerable to attack.

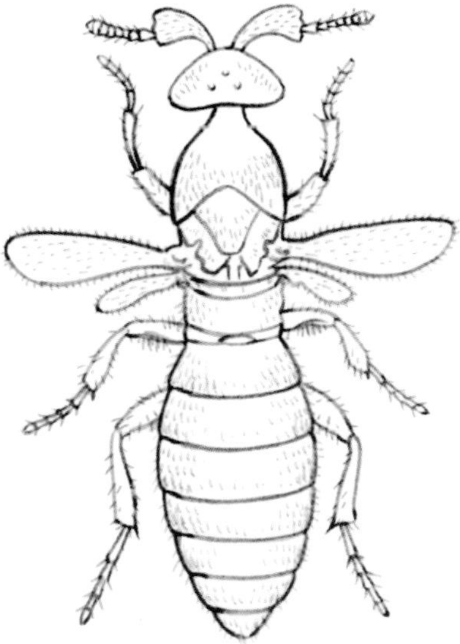

Male
Normal form

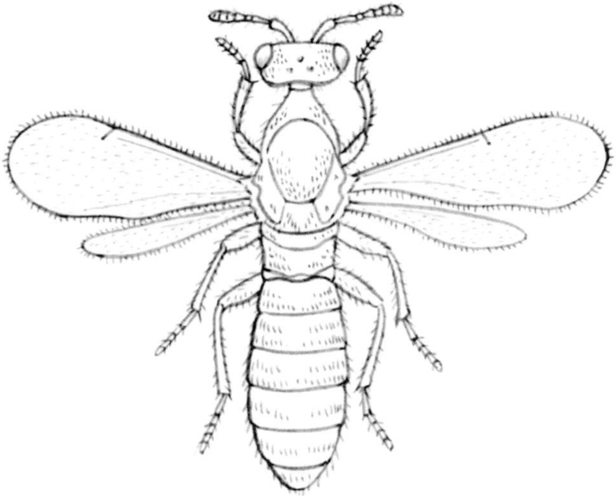

Female
Normal form

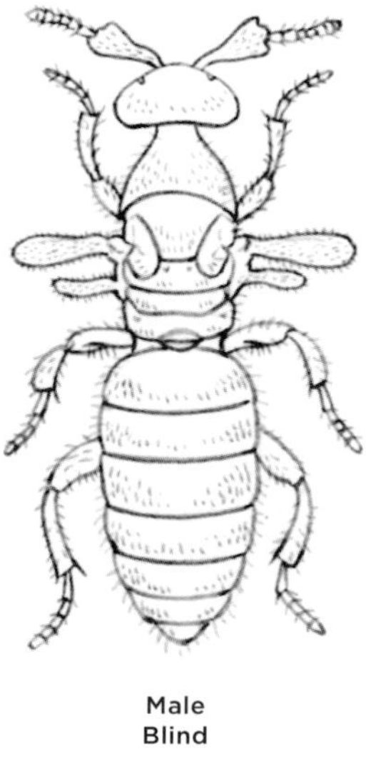

Male
Blind

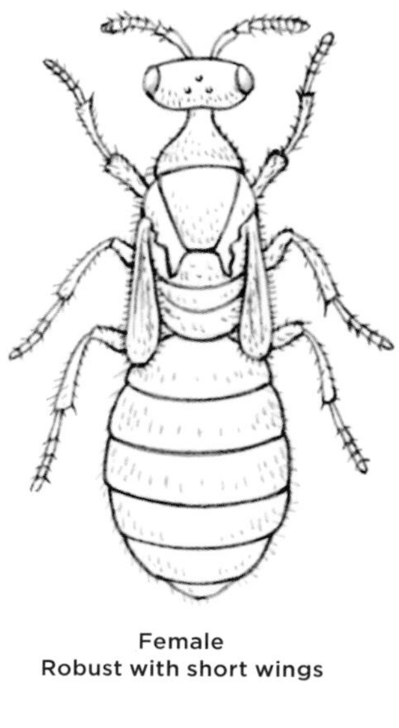

Female
Robust with short wings

10

Wasps and People
A Misunderstood Relationship

Allies, Outlaws, or Both?

Humanity has a great ambivalence towards wasps. Our patriarchal culture reveres the warrior image of social wasps, conveniently ignoring the fact that the wasps we want to emulate are all female, the Green Hornet should have been a female superhero. Meanwhile, we loathe yellowjackets for exploiting our urban and suburban lifestyles.

Much of our misery is of our own making. Our architecture invites wasps to nest there and our desire for plants and products from faraway lands comes with the added tax of invasive species that hitchhike on those shipments. Our careless and slovenly habits are rewarded with painful stings.

Our uneasy balance between respect for wasps—especially social species—and the sheer terror they present as stinging animals, is reflected in cultures the world over. Folklore and superstition abound, and wasps figure symbolically in art, literature, and entertainment. Wasps embody the best of us, and the worst of us, in character and image. The proverb "stirring up a hornet's nest" refers to our impression of social wasps as eternally angry.

Many positives have come from the study of wasps, though. We have sent wasps into space, for science, and they have inspired inventions, or improved existing products; wasps even provide cornerstones in the fields of ethology (the study of animal behavior) and chemical ecology. We employ many species in the biological control of pest insects, lessening the need for toxic pesticides in agriculture and our own vegetable gardens. We are indeed finally recognizing the importance of wasps to the maintenance and sustainability of natural and manmade ecosystems.

Still, phobias and hatred of wasps persist as an ever-present challenge to scientists and the public. Pest control providers stand to profit from a misunderstanding of wasp biology, and misidentification of the insect species concerned. Widespread sentiments of fear and loathing prevent consideration of precautionary measures, tolerance of ephemeral phenomena like solitary wasp aggregations, and alternative non-toxic approaches to control.

The most enlightened citizens are rebelling against the old standards of intolerance and embracing wasps in ways that were unimaginable only a decade or so ago. Homeowners are replacing lawns with native trees, shrubs, and herbs that better support native wildlife, from birds to bees. They increasingly resist the urge to intervene when an aphid outbreak occurs, understanding that parasitoid wasps, and various predators will likely cure the problem. They erect "bee condos" used by native, solitary bees, and also by solitary wasps that are preying upon garden pests.

The relationships between people and wasps are likely to become much friendlier in the future, provided we listen to, and trust, scientists. There is no substitute for dedicated study in illuminating the importance of other creatures to our human world.

Learning to Love Wasps
The public image of wasps is changing for the better with more education. Some people embrace this more enthusiastically than others, buying or building "insect hotels" that solitary parasitoid wasps also use.

Wasps in Religion and Mythology

Reverence and supernatural beliefs surrounding wasps abound, reflecting again our polar opposite attitudes toward them. Warfare, magical powers, and fertility appear as recurring themes.

Shamanistic cultures traditionally emphasize vertebrate animals, but yellowjackets, hornets, and other wasps figure prominently. Shapeshifting is typical: the soul of a Siberian shaman occasionally manifests in the form of a wasp, while a Mongolian shaman may choose a wasp as the costume for his external soul.

One elaborate legend comes from the Warao, or "boat people," indigenous Amerindians living mostly in the Orinoco Delta of Venezuela. Their mythical house of the swallow-tailed kite bears the nest of a honey wasp, *Brachygastra lecheguana*, and the testis of a human male. The Warao name for wasp is *ono*, for testis, acknowledging the fertility of the insect. The wasp nest and testis share similarities in color, shape, and internal structure. The chambers of the testis are separated by tissues suggestive of the layered combs in the hive. Honey wasps have multiple reproductive females distributed in different sections of the nest; likewise, the house of the swallow-tailed kite contains five deities: *Kanobo* Mawari, the Creator Bird and patron of light-shamans, the Black Bee (*Trigona hyalinata*), Blue Bee (*Trigona capitata*), Termite (*Nasutitermes corniger*), and a fierce Social Wasp (*Stenopolybia fulvofasciata*). Resident male spirits of the house, who are deceased light-shamans, gather around a board upon which the insect players interact to dictate the fate of life on earth.

Vanquishing an adversary with the aid of wasps is a common thread in mythology. The *Holy Bible* mentions hornets three times, each an instance in which the insects were enlisted to drive out enemies of the children of Israel. Seeking the powers of wasps, warriors of several North American tribes would burn wasps, gather the ashes, and apply them to their faces in the belief this would elevate their aggressiveness and enhance their battle skills. How did wasps

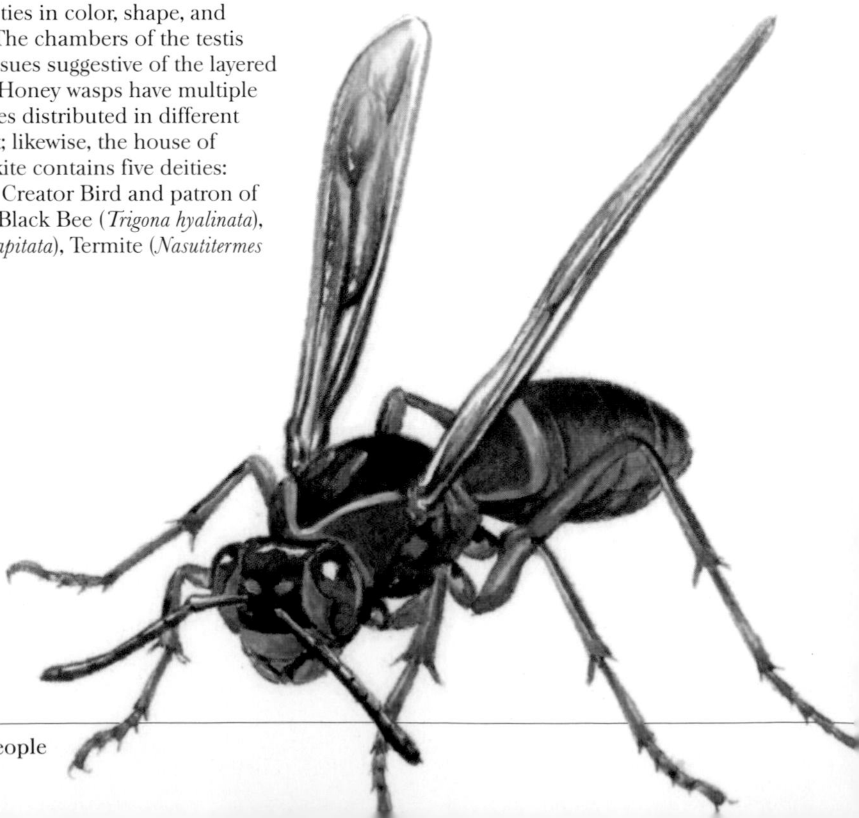

Able Protectors
A paper wasp (right and far right) is one of the guardians of the four cardinal points in Warao (Orinoco Delta in Venezuela).

Biblical Figure
Hornets (below) are mentioned in the Bible as allies in the protection of ancient Israel, driving out human adversaries.

"Boat People"
The Warao, indigenous people of the Orinoco Delta in Venezuela (right), have elaborate myths centered around the honey wasp and swallow-tailed kite.

come to sting in the first place? According to a legend of the Yurok of California, U.S.A., wasps acquired their venom by consuming the embers of a mythical burning arrowhead. Wasps and fire seem to go together.

The wasp was a sacred symbol of the Pharaohs, and an occasional alternative translation for "bee" in Egyptian hieroglyphics. Among the Maya of Central America, the wasp is likewise held in high esteem. One Mayan name for the planet Venus is *Xux Ek*, meaning "wasp star." Venus is celebrated in that culture as a powerful, sacrificial warrior, not unlike members of a wasp colony. The Hopi nation of the southwestern U.S.A. includes a hornet, or *Tatangaya*, among its kachinas, supernatural spirit beings that guide tribal activities.

Wasps are frequently portrayed as weapons. *Popol Vuh* (alternatively, *Popul Vuh*), the Mayan book of the dawn of life, includes a legendary episode in which the K'iche' achieved victory in a war by counseling with naguals, humans with the ability to shapeshift into animal forms. The naguals instructed them to load large vessels with hornets, small wasps, beetles, and snakes. The containers were deployed among the enemy, and the erupting organisms caused their adversaries to take flight themselves. The *Popol Vuh* also chronicles an incident where painted images of wasps on a blanket came alive and stung the enemy.

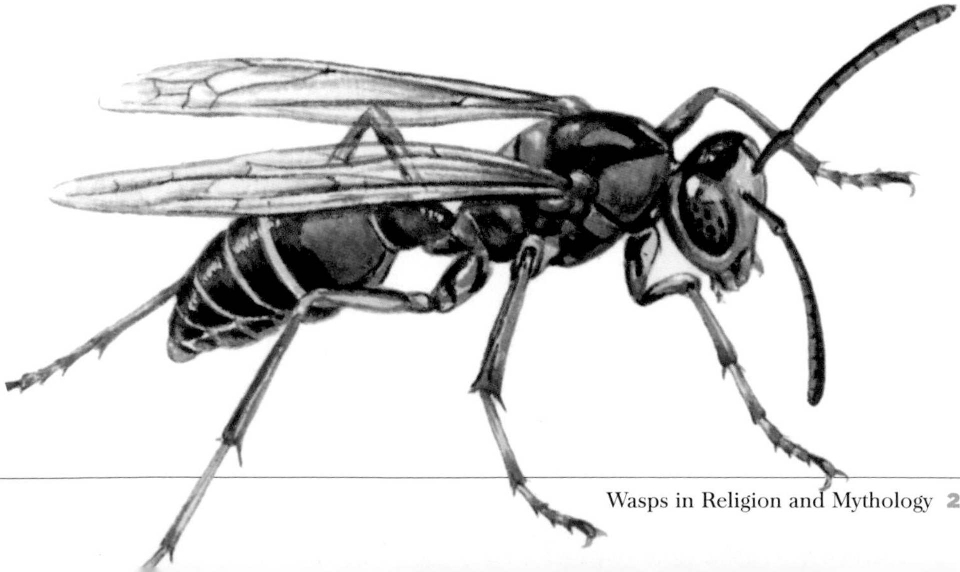

Wasps in Folklore and Superstition

A multitude of proverbs and fables have woven wasps into the fabric of past and present cultures. Exaggeration of the abilities of wasps, and their roles as omens of good or bad circumstances, are typical of these tall tales and folk sayings.

Superstitious Galls
European folklore treated "oak apples," a type of gall, much like a fortune cookie. Break one open, and the type of occupying insect would indicate good, bad, or ugly future events.

"Dirt Dauber"
Mud daubers, genus *Sceliphron* (right), are industrious masons, but a southern U.S.A. folktale is not as flattering, painting them as foolish know-it-alls.

Some indigenous peoples of South America assert that mankind learned pottery-making and homebuilding skills from wasps. A folktale from the southern U.S.A. begs to differ. The story goes that a hornet happened upon a "dirt dauber" constructing its nest and offered to show the slender wasp "how it is done." The mud dauber refused. Later, when time came for the solitary mason to add the door, the hornet again volunteered its wisdom. The mud dauber declined, went inside her nest, and affixed the door such that it sealed her within. The epithet of "dirt dauber" thus insinuates that a person is a know-it-all *and* ignorant.

The idea that wasps bring good fortune dates back millennia. One superstition originating from at least the first century AD asserts that if you desire good luck and immunity from enemies, you must kill the first wasp of spring. The insects fair better in other sayings. If a wasp flies into the house, good luck follows, goes one proverb. Abandoned nests of social wasps are common fixtures hung inside rustic cabins of Arkansas, Missouri, and other rural regions of the southern U.S.A., to evoke good tidings for the household. Young women in the Ozarks once pinned smaller nests to their undergarments to make themselves more desirable to men. Fertility symbolism again.

A European superstition holds that if an oak apple gall is opened, there are three possible occupants inside, each one symbolic. A fly inside indicates want, a worm (larva) predicts plenty, and a spider foretells death. From Pliny in the fifth century BC through at least the mid-1800s in London, galls have been prominent ingredients in folk medicine to cure various afflictions.

Wasps have been a predictor of weather in some corners. Consider the American folk saying "When hornets build nests near the ground, a harsh winter is expected." Another assertion from the rural southeast U.S.A. insists that if mud daubers build their nests low, it will be a dry year; if they build high, then expect heavier rains than average.

Velvet ants have their own mythology. Larger species are referred to as "cow killers" or "mule killers" in various parts of the U.S.A., owing to the excruciating sting of the female wasp. The Cherokee words for them are *da sûn tali atatsû ski*, or "stinging ant," and, less frequently, *nun yunú wi* ("stone dress"), for the insect's dense exoskeleton.

Paper Nest
The nests of paper wasps are staple décor in nature centers and rural cabins, but they also symbolize fertility. Young women in the Ozarks of the U.S.A. once pinned them to their undergarments.

Wasps in History, Tradition, and Pop Culture

The line between legend and history is often blurred in wasp lore, and some aspects of culture bleed into the entertainment category. In any event, we reveal much about ourselves in how we relate to wasps, be it positive, negative, or neutral.

Eating Wasps
The larvae and pupae of wasps is part of the human diet in many Asian cultures. Eating insects is called entomophagy, and it is getting serious scientific scrutiny as a sustainable food source.

One intriguing aspect of our attitudes is the use of acronyms. "W.A.S.P." traditionally translates to White Anglo-Saxon Protestant. It has become a certifiably derogatory cultural reference, at least in the U.S.A. where it refers principally to excessively privileged people of British descent who achieve economic and social success largely through nepotism and marriage within their class. WASPs are the epitome of "establishment" politics.

A more complimentary use of W.A.S.P. is the Women Airforce Service Pilots. The organization formed in 1943 as a civilian group of pilots who tested aircraft, trained other pilots, and ferried airplanes for the U.S. Army Air Forces, briefly during Second World War. They were female, they flew…waspish, indeed. A comprehensive list of other definitions found in the online free dictionary runs the gamut from World Association for Social Psychiatry to the band We Are Sexual Perverts—however, the latter may well

Sweet Scooter
Italian Vespa scooters are one of the most popular modes of transportation in many European cities. Their sleek style and pleasant hum are an admirable tribute to the wasps for which they are named.

be just a popular myth. Considering that real wasps exhibit parthenogenesis, incest, and other deviant reproductive behavior, that seems fitting.

The practice of human consumption of insects is called entomophagy. Historically, collecting nests of social wasps for the larvae and pupae has been documented on all habitable continents except Europe, as well as islands in Oceania. Today, this is uncommon. Japan's history of wasp-eating probably predates the first documentation in 1715, but it persists in contemporary times. Villagers in Kushihara (now part of Ena) in Japan collect small subterranean nests of yellowjackets, *Vespula flaviceps*, and "grow" them inside special huts. The wasps are provided with chunks of raw meat, and the nests allowed to expand. In autumn they harvest the juicy larvae and pupae. There is even a wasp festival, *hebo matsuri*, held annually on 3 November. The brood of several species of hornets (*Vespa*), the night hornet *Provespa barthelemyi*, and paper wasp *Polistes sagittarius*, are eaten in China. Other social wasps are part of the diet in Laos and Thailand, and Kalimantan (Borneo, Indonesia).

Mascots bridge pop culture and entertainment, but are serious business. On 3 April, 1998, Georgia Institute of Technology (Georgia Tech) (U.S.A.) filed suit against the Salt Lake Buzz minor league baseball team over perceived trademark infringement of its mascot, "Buzz" the yellowjacket. In October, 2001, and a countersuit or so later, the Buzz changed its name to "Bees" and settled for $600,000 in damages. Georgia Tech spent over $700,000 in litigation expenses. So far, the Charlotte Hornets of the National Basketball Association have escaped such court proceedings.

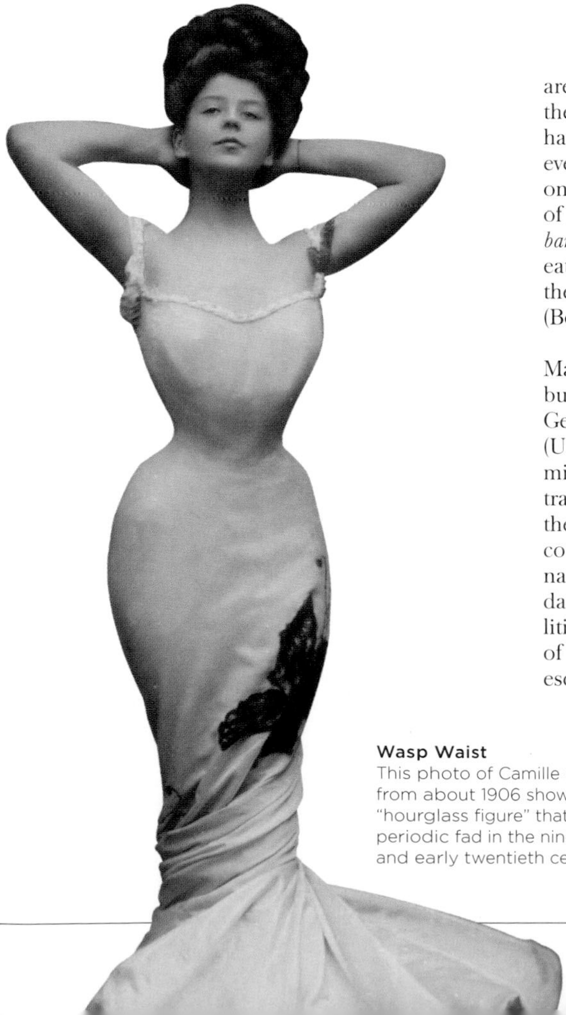

Wasp Waist
This photo of Camille Clifford from about 1906 shows the "hourglass figure" that was a periodic fad in the nineteenth and early twentieth century.

Wasps in Literature, Art, and Entertainment

The fictional representation of wasps follows in the tradition of good versus evil in our appraisal of wasps. Realistic renderings and descriptions temper the extremes, and are wonderful vehicles for engendering fascination.

Jean Henri Fabre
The celebrated "father of entomology" (above), studied many types of insects in his native France. He published prolifically throughout the late 1800s and early 1900s, for both the scientific community and the general public.

One of the earliest, most exquisite artistic pieces depicting wasps is the so-called "Cretan Hornet," a gold pendant dating between 2000–1700 BC. It comes from Chrysolakkos necropolis, a cemetery in Mallia, and is of Minoan origin. The twin, mirrored insect subjects are likely not hornets, nor bees as some scholars have claimed. Instead they probably portray paper wasps (*Polistes* sp.). Today, this artifact is housed in the archaeological museum at Heraklion, Crete.

In the late 1600s, scientific illustrator Maria Sibylla Merian painted relationships between organisms, especially caterpillars and their host plants. She laid the foundation for male naturalists who came after her. John James Audubon included wasps in his sketchbooks, and in "bird dramas" that had the insects evading avian predators. Philip Henry Gosse's unpublished manuscript *Entomologia Terrae Novae*, from 1833, features stunning illustrations of insects from Newfoundland, Canada. Gosse was an equally talented writer.

More recent masters of insect art include E.A. Seguy, a French designer of the 1920s, who rendered excellent paintings of preserved specimens and also translated them into stylized motifs. Walter Linsenmaier (1917–2000) was a Swiss entomologist and artist known for portraying insects in action, in their natural habitats. French painter Bernard Durin (1940–1988) created hyper-realistic portraits of insect subjects that almost crawled off the canvas.

Jean Henri Fabre, the celebrated French entomologist, was arguably the first scientist to make the insect world publicly accessible and captivating. His lively prose made liberal use of anthropomorphism, but his contagious enthusiasm and gift for suspense remain nearly unparalleled.

Censored Story
An episode of Lewis Carroll's *Through the Looking-Glass* never made it into the final book because the illustrator refused to render it. The vignette featured Alice's encounter with an elderly male hornet.

Lewis Carroll
The Wasp in a Wig
A "Suppressed" Episode of
Through the Looking-Glass
and What Alice Found There
With a Preface, Introduction and Notes by
Martin Gardner
Lewis Carroll
THE WASP IN A WIG

"The Wasp in a Wig" was a suppressed story from Lewis Carroll's *Through the Looking-Glass*, unearthed in 1977, 107 years after it was penned. Artist John Tenniel refused to illustrate it, and though four other artists volunteered, it was never published. The episode is an encounter between Alice and an elderly, irascible male vespid, ungrateful, initially, for her attempts to comfort him.

Music displays only minor affinity for wasps, though bees and ants are more prominent. Entomologist and audiophile Joseph Coelho compiled a thorough record of insects in the rock and roll genre up through the year 2000, including band names, albums, and individual tracks.

Gary Larson began his career as an entomologist before finding his true calling as the cartoonist behind The *Far Side*. He frequently humanized insects, an artistic category entomologist Charles L. Hogue coined as "bugfolk," a recurring style throughout art history. Larson has since spawned many imitators, several quite creative and successful in their own right.

Fashion Statement
"What the..." "I don't care if you think it's weird! Stripes are overrated anyway." Canadian Devon Henderson's anthropomorphic wasps are a contemporary example of "bugfolk." Devon does scientific illustration, too.

....WHAT......
...THE....
I don't care if you think it's weird! Stripes are overrated anyway.
DH.

Wasps as Pests and Invasive Species

While the benefits of wasps are many, there is no denying their negative economic, physical, and psychological impacts. Those effects can be mitigated or prevented, if we have the will.

End Game
A yellowjacket trap with its mass casualties. Baited with a special attractant chemical, the device kills few other insects besides scavenging social wasps of the genus *Vespula*.

Pest is how we describe any organism we perceive as a competitor for our resources, a danger to our health, or that compromises the interests of our economy. We conflate "nuisance" with pest, and treat the species with weapons out of proportion to its damage. Further, our agricultural practices of industrial-scale monoculture magnify the effects of pests. Our quest for exotic landscape plants, and products from overseas, compound matters by introducing foreign species.

Grain crops and nut orchards are particularly susceptible to attack by stem sawflies (Cephidae) and seed chalcids (Eulophidae and Torymidae), while logging operations battle horntail wood-wasps (Siricidae). Human laborers risk stings when ripening fruit is swarmed by yellowjackets and paper wasps, and local economies dependent on tourism and outdoor recreation face constant threat from social wasps that harass campers, picnickers, and other vacationers. Yellowjackets have periodic eruptions in their populations, reaching densities that may exceed our ability to overcome.

Stings are, normally, a minor inconvenience of localized pain, swelling, and itch that alleviates within hours. A sting to the tongue or throat can be another matter, as swelling can constrict breathing passages and lead to suffocation. People with hypersensitive immune systems are at even greater risk. Allergies to insect venoms mean that any sting can be life-threatening, sending the victim into anaphylaxis. Self-injection devices that deliver life-saving epinephrine (adrenalin) are available, but often cost-prohibitive.

Watch Out!
Wasps are fond of sweets, so never serve canned beverages at your outdoor meals. Swelling of the tongue or throat from a sting can be life-threatening.

Pest Control
A pest control technician (above) extracts a nest of yellowjackets from between the walls of a house. Professional intervention is necessary for such situations, but outside the home wasps should be respected, welcome insects.

Wasps also exert profound mental distress. Spheksophobia is the intense fear of wasps; a subcategory of entomophobia, the paralyzing fear of insects in general. Attaching stigma to a person so afflicted does not help. Thankfully, professional counseling and desensitization techniques often result in resumption of normal daily activities for those who seek a solution.

Many wasp species have expanded their geographic ranges to new continents with the help of global commerce. It need not be the price of international trade, but we are so far unwilling to absorb the costs of thorough port inspections and other preventive measures. Most recently, a colony of the Asian giant hornet, *Vespa mandarinia*, was discovered and destroyed in Nanaimo, British Columbia, Canada in September, 2019. A single, dead specimen, not associated with the Canadian nest, turned up near Blaine, Washington, U.S.A. in December. Dubbed "murder hornet" by the media, the species continues to appear. A live specimen was spotted in Langley, BC on 15 May, 2020, and a road-killed specimen was found near Custer, Washington. Will it become an invasive threat to beekeeping in North America? Stay tuned.

Invasive
The European paper wasp, *Polistes dominula* (left), is one of numerous wasp species that have escaped their native lands to wreak havoc on foreign ecosystems. They become pests of people, crops, and forests, and out-compete native species.

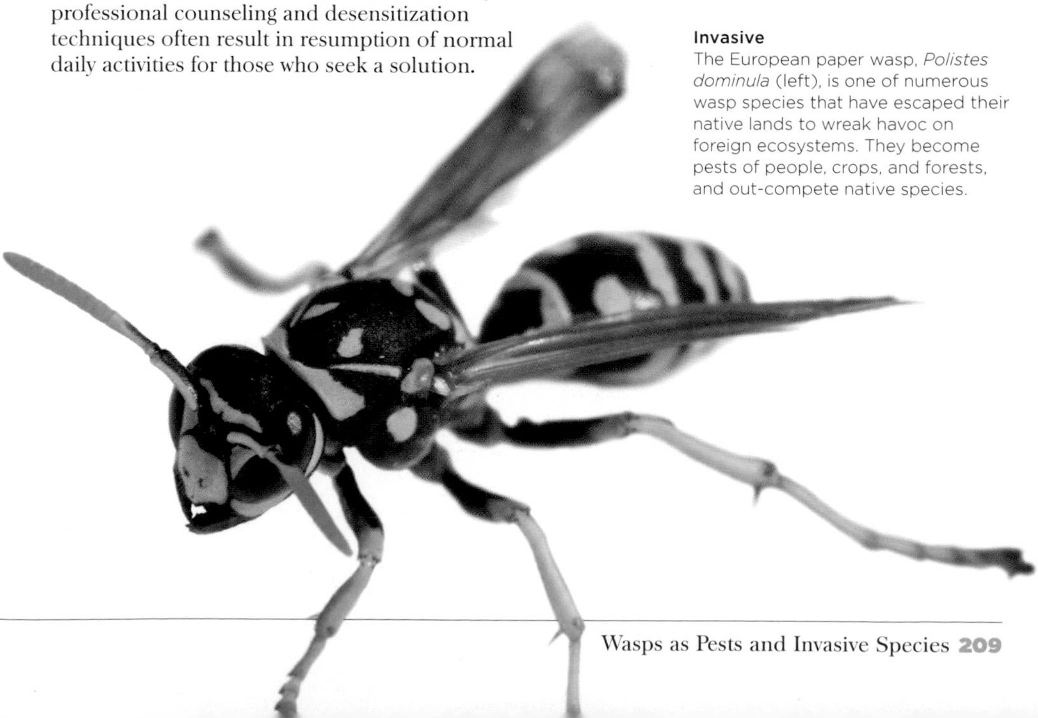

Wasps in Science and Invention

Wasps have played a critical role in the advancement of human civilization through the inspiration and improvement of inventions, as subjects of scientific study, and tools of pest control and medicine.

Wasps on Board
At least one American space mission included a colony of live hornets as part of a science experiment in 1992. The mission was, unfortunately, a failure, as the wasps perished.

Cai Lun is widely regarded as the father of paper making, but it is doubtful that he was inspired by wasps. There are records of paper use in China that predate him, and paper in Lun's day consisted of recycled rags of hemp, linen, or cotton, plus other plant fibers. Fast forward to the sixteenth century when a dearth of cotton and linen resulted in a paper shortage. In 1719, French naturalist Antoine Ferchault de Réaumer, who probably did observe the activities of wasps, is credited for updating the manufacturing process to use wood pulp as the raw material.

Oak galls, rich in tannins, lend themselves to several uses. Iron gall ink combined iron salts with tannic acid, plus a binding agent (usually gum arabic) to create pigment. Gall-derived inks have been utilized since the fifth century, right through most of the nineteenth. Allepo galls, found in eastern Europe and western Asia, were used by the Greeks in dyeing wool; and Pliny mentioned their use as a hair dye. Somali women use local galls for tattooing ink.

Allies of Agriculture
A technician at an insectary (left) processes cards containing the wasp-parasitized eggs of a crop pest, to be released into fields for biocontrol. Some of these tiny parasitoids are available in retail outlets.

Paper Inspiration
A Bald-faced Hornet, *Dolichovespula maculata* (below), scrapes wood fibers she will chew into pulp for her colony's nest. Observations of paper wasps probably led to using wood pulp in paper production as we know it today.

Before he became the celebrated father of research on human sexuality, Alfred Charles Kinsey was known for his investigations into the classification of gall wasps in the 1930s. He applied a taxonomic approach to re-classifying sexual "perversions" as behaviors well within the bounds of biological, if not moral, normalcy.

Galls, wasp nests, and wasps themselves were consistent ingredients in folk remedies but, mercifully, medicine has progressed. Cleopatra allegedly tested wasp venom on prisoners, but the science of chemical ecology has a more humane agenda. Various compounds isolated from wasp venoms show great pharmacological potential. A chemical in the venom of a Brazilian social wasp, *Polybia paulista*, targets cancer cells by recognizing anomalies in the cell membrane and eroding a gaping hole.

Niko Tinbergen, Karl von Frisch, and Konrad Lorenz shared the 1973 Nobel Prize in Physiology or Medicine for their landmark studies of animal behavior, creating the scientific discipline of ethology. Tinbergen's principal subject was the European beewolf, *Philanthus triangulum*. Ironically, von Frisch studied its prey, the honey bee. Laboratory wasp research has included a failed Japanese experiment with oriental hornets aboard the *Endeavour* space shuttle in 1992.

Wasps can even help solve crimes. Take a murder case in Tennessee, U.S.A. where a skull was discovered with a paper wasp nest inside. Forensic detectives worked backward to determine time of death, which coincided with the victim's initial disappearance.

Welcoming Wasps into Your Outdoor Space

In *Wasp Farm*, Howard E. Evans includes a chapter titled "How to Attract Wasps and Why." It is a testament to the author's sense of humor and his dedication to promoting the welfare of our stinging insect friends.

Housing Crisis
Solitary wasps face a real estate shortage, so putting up artificial abodes is helpful. Here, a male keyhole wasp, *Trypoxylon* sp., peeks from its mate's nest in a cardboard tube.

There is less sweat involved in modifying your yard than there is changing the landscape of your mind. Doing research and other homework *before* you act ensures better results than impulsively reaching for a canister of insecticide or yanking an unfamiliar flower. Consult your local chapter of the Native Plant Society, the Xerces Society, and other conservation organizations for help. Cultivate good habits as diligently as you cultivate the garden. Resist expectations imposed externally by the commercial landscape and nursery industry.

"Weed tolerance" is a simple way to begin. Noxious weeds, as listed by local, state, provincial, and national government agencies, have no place in the garden and should be eliminated. The remaining plants that volunteer will be mostly natives that should be embraced.

Raising your threshold for insect damage is also helpful. We severely underestimate the ability of plants to defend themselves, their ability to summon parasitoids to their rescue, and to recover from insect herbivory. Native, local plants far surpass exotic cultivars in each of these regards. We are also ignorant of the ephemeral nature of most insect attacks.

Wasps face a housing shortage, too. You may wish to purchase or construct a "bee condo." The best versions are simply a block of wood with drilled holes of varying diameters, or bundles of hollow twigs. Placed at least a meter off the ground and facing south or east, such devices attract both solitary bees that pollinate your garden, and solitary stinging parasitoids that rid your yard of unwanted caterpillars and other pesky insects.

Admittedly, discouraging social wasps is challenging. Paper bag facsimiles of yellowjacket nests, suspended from under an eave, are of dubious effectiveness in repelling real wasps. The same is true for painting the underside of the porch sky blue to deter nest-building. It does not usually work. Rockwork and retaining walls attract queens looking to nest, so reconsider those landscaping choices. Reduce the likelihood of stings at the barbecue by serving beverages in clear glassware rather than cans and opaque bottles, lest someone unwittingly gulp down a wasp.

Douglas Tallamy, in his books *Bringing Nature Home* and *Nature's Best Hope*, advocates the rewilding of the world through individual action by private landowners. Replacing lawns with more natural meadows, and planting indigenous trees, shrubs, vines, and herbs are part of his recipe for converting urban and suburban parcels into linked refuges and corridors for wildlife, wasps included.

A Welcome Visitor
A mud dauber, *Sceliphron* sp. (below), drinks nectar from a flower. Wasps may or may not pollinate flowers, but they service your garden anyway by controlling other insects that could become pests.

Wasp Feeder?
Worker Southern Yellowjackets, *Vespula squamosa* of the U.S.A., congregate on a hummingbird feeder (above). They may be a nuisance at times, but they also prey on pest insects that would otherwise overwhelm your garden.

11

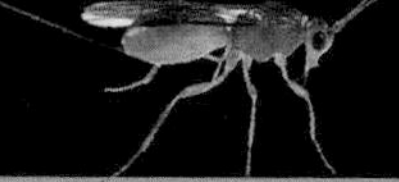

A Wasp Family Album
Microcosm of Diversity

From the Familiar to the Fascinating

The following is not intended to be a field guide to the wasps of the world. It is by no means comprehensive, but the families of wasps included here are those most likely to be encountered, those most important economically or ecologically, and those not covered elsewhere in the text. They are presented here in taxonomic order, from "primitive" sawflies to "advanced" social wasps. Identification of wasps, even to family-level, is rarely accomplished by comparison to images alone. Most professional entomologists who are wasp experts specialize in one family.

There are currently about 83 families of wasps recognized, but this is likely to change. The family Trachypetidae was erected in 2020, for example. Its members were formerly considered part of the Braconidae. As scientists achieve better understanding of the evolutionary relationships between wasp families, there is continual "lumping" of two or more families into one, or "splitting" of families from one into two or more. This is the essence of taxonomy, the ever-changing classification of living things.

Given the less-than-absolute nature of classification, many of the numbers for genera and species in the following tables are compiled from several sources, many of which are estimated. A "~" symbol stands for "approximately," for example, and ">" means "more than." What is true today may not be true tomorrow, of course.

Insects are not like birds. There are more species, by several orders of magnitude. The observable characters of birds are generally consistent, and useful for identification. Not so for wasps. Mimicry complicates matters to the degree that it may be nearly impossible to tell if what you are looking at is a wasp, a fly, a moth, or some other insect. Do not be discouraged from "wasp-watching" by the steep learning curve, though. Find an entomologist mentor and/or consult some of the additional references in the back of this book. Have fun and record your experiences. You could very well contribute something new to our collective knowledge. We know *that* little.

Family Xyelidae

FAMILY	Xyelidae
OTHER NAMES	Xyelid sawflies
DISTRIBUTION/WHERE FOUND	Holarctic—across entire northern hemisphere (5 genera, 63 species)
SOLITARY OR SOCIAL?	Solitary
IDENTIFICATION	Small (0.1–0.15 inches, mostly 0.1–0.15 inches (3–15mm/3–4mm)); antennae distinctive: first segment of flagellum the longest, thickest; ovipositor of female obvious
SIMILAR WASPS	Braconidae, Ichneumonidae
HABITAT	Forests
NEST	None
IMPORTANCE	No economic importance

Xyelids are considered the most primitive wasps, and there are far more extinct genera than extant ones. Living species are restricted overwhelmingly to temperate climates, with one species in the neotropics. Larvae feed within pollen cones of pine, inside the buds and shoots of firs, or externally on foliage of deciduous trees. The adults are seldom seen, but may congregate around pollen cones of pines, or willow catkins in early spring.

Family Pamphiliidae

FAMILY	Pamphiliidae
OTHER NAMES	Pamphiliids, web-spinning sawflies, leaf-rolling sawflies
DISTRIBUTION/WHERE FOUND	Temperate North America and Eurasia (13 genera, 297 species)
SOLITARY OR SOCIAL?	Solitary, but some species are gregarious as larvae
IDENTIFICATION	Often colorful; large, flat, round or square head; long, very slender antennae; net-like wing venation
SIMILAR WASPS	Other sawfly families
HABITAT	Mostly forests, but also on urban and suburban landscape trees
NEST	None, but larvae may construct communal silk shelters
IMPORTANCE	A few species are occasional forest pests

Pamphiliids are not well-studied, despite being attractive and reasonably common insects. They are seldom found far from their host trees, especially in coniferous forests. It takes close examination of boughs to find them, and the larvae may be more conspicuous given that some species spin extensive webbing laden with frass (insect feces) and fragments of plant matter.

Family Argidae

FAMILY	Argidae
OTHER NAMES	Argid sawflies
DISTRIBUTION/WHERE FOUND	All continents but Antarctica (60 genera, 913 species)
SOLITARY OR SOCIAL?	Solitary, though larvae may feed gregariously
IDENTIFICATION	Small (0.1–0.4 inches (3-10mm)); third segment of antenna (flagellum) undivided into flagellomeres, but may be forked or otherwise modified in males; often colored black and red or orange
SIMILAR WASPS	Tenthredinidae, Pergidae, Diprionidae, Crabronidae
HABITAT	Found in a variety of habitats, including urban lots
NEST	None
IMPORTANCE	Some species are considered pests of trees, shrubs, or herbs

Argids are among our most commonly-encountered sawflies, especially on umbelliferous flowers. The caterpillar-like larvae feed externally, and often gregariously, on the foliage of many types of plants. The purslane sawfly, *Schizocerella pilicornis*, mines between the layers of the leaves of its host plant. Two species have found employment as biocontrol agents for noxious weeds. Larvae of a few species spin communal cocoons; and females of other species guard their eggs and young larvae.

Family Cimbicidae

FAMILY	Cimbicidae
OTHER NAMES	Club-horned sawflies, giant sawflies, cimbicid sawflies
DISTRIBUTION/WHERE FOUND	Holarctic—across entire northern hemisphere, plus South America (16 genera, 182 species)
SOLITARY OR SOCIAL?	Solitary
IDENTIFICATION	Large (up to 1.2 inches (30mm)), robust, with clubbed antennae
SIMILAR WASPS	More likely to be mistaken for large bees or hornets
HABITAT	Mostly temperate forests, riparian wetlands
NEST	None
IMPORTANCE	Rarely pests of deciduous trees

Cimbicids can be intimidating insects, flying with a loud, droning buzz. Males may lay claim to territories that include landmarks such as isolated shrubs. From those vantage points they give chase to passing females or rival males. Males may have enlarged and modified hind legs. Both sexes have large jaws. They occasionally girdle the twigs of living trees while gnawing on bark. Larvae are frequently mistaken for large caterpillars.

Family Diprionidae

FAMILY	Diprionidae
OTHER NAMES	Diprionids, conifer sawflies
DISTRIBUTION/WHERE FOUND	Holarctic—across entire northern hemisphere; adventive (non-native) in Australia (11genera, 136 species)
SOLITARY OR SOCIAL?	Solitary, but larvae often feed gregariously
IDENTIFICATION	Small (0.23–0.47 inches average (6–12mm); compact antennae usually saw-like (serrate) in females and comb-like (pectinate) in males; rarely found away from host trees
SIMILAR WASPS	Other sawflies
HABITAT	Coniferous forests, but even on ornamental conifers in urban landscapes
NEST	None
IMPORTANCE	European pine sawfly, *Diprion similis*, is an invasive species; prone to epic population outbreaks

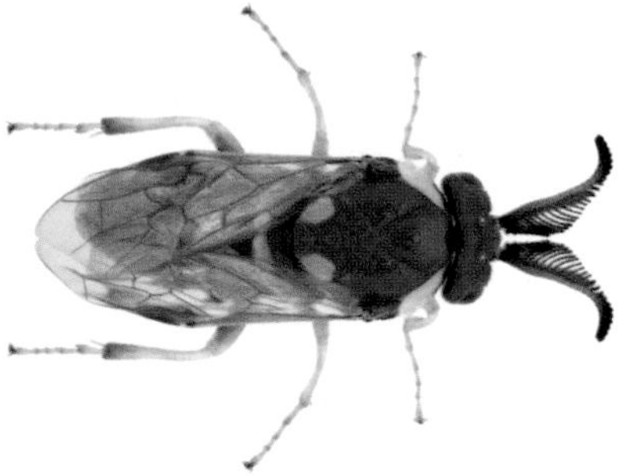

Conifer sawflies feed exclusively on conifers as larvae, and are frequent forest pests. They usually weaken trees and slow their growth, rather than killing them outright. Despite rigorous study, scientists are still having difficulty in defining the species.

Family Pergidae

FAMILY	Pergidae
OTHER NAMES	Pergid sawflies
DISTRIBUTION/WHERE FOUND	Australia, South America, Eastern North America; adventive (non-native) in South Africa (60 genera, 442 species)
SOLITARY OR SOCIAL?	Solitary, but larvae of some are gregariously nomadic
IDENTIFICATION	Highly variable and defy easy identification
SIMILAR WASPS	Argidae; some pergids are excellent mimics of Vespidae
HABITAT	Variable
NEST	None
IMPORTANCE	Larvae of some species are toxic to livestock animals, often lethally so; some species are crop pests and/or invasive species

Pergids are the sawflies of Australia and South America, most diverse there, though the family has a wider distribution. The larvae have equally diverse feeding habits. Most species consume the leaves of deciduous trees, but larvae in the genus *Perreyia* roam in packs over the ground, eating dead or dying leaves. Semi-aquatic ferns are on the menu for *Warra froggatii*, and fungi are a recorded host for *Decameria rufiventris*. Females of some *Pseudoperga*, *Philomastix*, and *Cladomacra* stand guard over their eggs, and may protect the young larvae as well. *Cladomacra* are unusual in another regard: females can be wingless.

Family Tenthredinidae

FAMILY	Tenthredinidae
OTHER NAMES	Tenthredinids, common sawflies
DISTRIBUTION/WHERE FOUND	All continents but Antarctica (~400 genera, >5500 species)
SOLITARY OR SOCIAL?	Solitary, though larvae of some feed gregariously
IDENTIFICATION	Highly variable and defy easy identification
SIMILAR WASPS	Many species mimic Vespidae, Pompilidae, Crabronidae, Ichneumonidae, Braconidae
HABITAT	Variable; most diverse in forests and wetlands
NEST	None, though a few species are gall-makers
IMPORTANCE	Many are considered pests of trees, shrubs

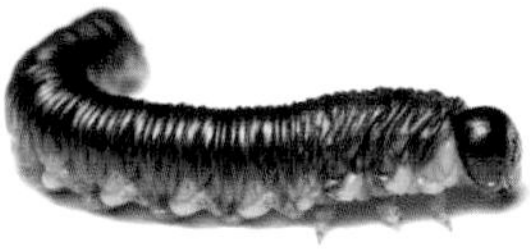

Common sawflies are indeed common, but many are outstanding mimics of stinging wasps and dismissed accordingly in the field. The larva stage can be a leaf-nibbler (most species), a gall-maker, leaf miner, or, rarely, a borer in stems. Most larvae resemble caterpillars, but some are covered in thick, waxy pruinosity or filaments. Some appear slimy and slug-like. Adult sawflies are most abundant and conspicuous in spring and early summer, on foliage in the forest understory, on flowers in meadows and fields, and in your garden. Many species are regarded as pests if their host plant is one that we covet ourselves.

Family Cephidae

FAMILY	Cephidae
OTHER NAMES	Cephids, stem sawflies
DISTRIBUTION/WHERE FOUND	Holarctic—across entire northern hemisphere (21 genera, 160 species)
SOLITARY OR SOCIAL?	Solitary
IDENTIFICATION	Small, slender, cigar-shaped
SIMILAR WASPS	Siricidae, Xiphydriidae, but much smaller; Braconidae, Ichneumonidae
HABITAT	Grasslands, meadows, forest edges
NEST	None, larvae are borers
IMPORTANCE	Some are crop pests, others a garden nuisance

Stem sawflies are not uncommon, but are rarely seen or recognized. Their extremely slender appearance also allows them to remain inconspicuous while perched head down on a stem or shoot. The larvae are borers inside grass stems, or in the stalks of woody plants. As such they are often considered pests.

Family Siricidae

FAMILY	Siricidae
OTHER NAMES	Siricids, horntails, horntail wood wasps
DISTRIBUTION/WHERE FOUND	Holarctic—across entire northern hemisphere; adventive (non-native) in New Zealand, Tasmania, South Africa, and South America (20 genera, 124 species)
SOLITARY OR SOCIAL?	Solitary
IDENTIFICATION	Large to very large; cigar-shaped body; square head; female with stout ovipositor and knob- or spear-like cornus
SIMILAR WASPS	Vespidae, Crabronidae
HABITAT	Forests
NEST	None, larvae bore in wood
IMPORTANCE	Some species are considered forest pests; a few are invasive species

You are not likely to ignore a horntail, as large and loud as they are. The larvae are wood-borers by proxy: the female wasp deposits a symbiotic fungus when she lays an egg. The fungus advances through the wood ahead of the larva, predigesting the cellulose. The horntail larva essentially feeds on the fungus. Foresters call these wasps by profane names related to the manner in which the female "humps" a tree while ovipositing. A number of siricids are invasive species, particularly in North America, South America, South Africa, and Australia.

Family Xiphydriidae

FAMILY	Xiphydriidae
OTHER NAMES	Xiphydriids, wood wasps
DISTRIBUTION/WHERE FOUND	Eastern North America, mountains of western North America, northern Europe, eastern Asia, Australia, New Zealand, and scattered locations in the tropical Americas. (28 genera, 146 species)
SOLITARY OR SOCIAL?	Solitary
IDENTIFICATION	Resemble small horntails, but with round head, not square
SIMILAR WASPS	Siricidae, Pamphiliidae, Crabronidae
HABITAT	Forests
NEST	None, larvae bore in wood
IMPORTANCE	No economic importance

Wood wasps are among the more scarce of all Hymenoptera, and consequently little is known about their hosts, diversity, and geographical distribution. They are most likely to be encountered, along with horntails and many wood-boring beetles, on freshly-felled timber, windfalls, and recently broken branches.

Family Orussidae

FAMILY	Orussidae
OTHER NAMES	Parasitic wood wasps, orussids
DISTRIBUTION/WHERE FOUND	Widely scattered over all continents but Antarctica (16 genera, 82 species)
SOLITARY OR SOCIAL?	Solitary
IDENTIFICATION	Small (0.2–0.5 inches (5–14mm)); globose head; distinctive antennae, like walking canes, set low on face; compact and cylindrical in body form
SIMILAR WASPS	Commonly mistaken for ants, but also Tenthredinidae, Crabronidae, Ichneumonidae
HABITAT	Uncommon to locally common in forests, riparian corridors, where wood-boring insect larvae are found
NEST	None
IMPORTANCE	No economic importance

Parasitic wood wasps are enigmatic for their anatomical resemblance to sawflies, but with a parasitoid lifestyle. Orussids are usually seen on logs or dead, standing trees where females search for the larvae of jewel beetles (Buprestidae), and perhaps other wood-boring insects. She deposits her egg in the host tunnel, not on the host itself. The larva that hatches follows the trail of the host's frass, subsisting on the poop until it reaches the source, then feeds as an ectoparasitoid. Adult orussids can jump, owing to a massive muscle that replaces one of the flight muscles.

Family Stephanidae

FAMILY	Stephanidae
OTHER NAMES	Stephanids, crown wasps
DISTRIBUTION/WHERE FOUND	All continents but Antarctica, most diverse in subtropics (11 genera, 342 species)
SOLITARY OR SOCIAL?	Solitary
IDENTIFICATION	Extremely slender body; spherical head with "crown" of teeth or tubercles; obvious "neck"; thickened and toothed femur on hind leg; females with long ovipositor
SIMILAR WASPS	Aulacidae, Ichneumonidae, Braconidae
HABITAT	Forests, riparian corridors, where wood-boring insect larvae are found
NEST	None
IMPORTANCE	No economic importance

Crown wasps are encountered infrequently in temperate regions. The larvae are external parasitoids of wood-boring beetles, mostly Cerambycidae (longhorned beetles) and Buprestidae (jewel beetles). At least one stephanid, *Schletterius cinctipes*, is a parasitoid of horntails, and has been imported to Tasmania as a biocontrol agent of the equally exotic *Sirex noctilio*. Stephanids are most diverse in the tropics.

Family Trigonalidae

FAMILY	Trigonalidae (sometimes "Trigonalyidae")
OTHER NAMES	Trigonalids
DISTRIBUTION/WHERE FOUND	The Americas, Europe, Asia, Australia (16 genera, 92 species)
SOLITARY OR SOCIAL?	Solitary
IDENTIFICATION	No characteristics easily work for field identification
SIMILAR WASPS	Vespidae, Crabronidae, Tenthredinidae, Ichneumonidae
HABITAT	Variable, most diverse in forest understory
NEST	None
IMPORTANCE	Rarely encountered, importance unknown

Trigonalids may be the most rarely-seen of all wasps. Their life cycles defy logic. Females deposit hundreds, if not thousands, of minute, hard-shelled eggs on the underside of leaves. The eggs are built to withstand a chomping caterpillar or sawfly larva and once ingested, the eggs hatch. Each trigonalid larva then begins seeking another parasitoid inside the larva. Should it find none, it must wait for one to appear. This hyperparasitoid lifestyle is apparently successful often enough. Stranger yet are trigonalids that must wait for the caterpillar to be chewed up and fed to a yellowjacket larva, the intended host reached in a most Rube Goldberg way.

Family Aulacidae

FAMILY	Aulacidae
OTHER NAMES	Aulacids
DISTRIBUTION/WHERE FOUND	All continents but Antarctica (2genera, 185 species)
SOLITARY OR SOCIAL?	Solitary
IDENTIFICATION	Abdomen attached high on thorax; ovipositor often downcurved near the tip
SIMILAR WASPS	Ichneumonidae, Braconidae, Gasteruptiidae
HABITAT	Forests and forest edges where wood-boring insect larvae are found
NEST	None
IMPORTANCE	No economic importance

Aulacids are probably more common than suspected, but their resemblance to ichneumon wasps and their mimicry of some stinging parasitoids is no doubt responsible for misidentification in the field. The larvae are parasitoids of wood-boring jewel beetles (family Buprestidae), longhorned beetles (Cerambycidae), and wood wasps (Xiphydriidae). One is therefore most likely to find them at recently felled trees in logging camps and sawmills, or in the immediate wake of forest fires. Grooves on the inside of the coxae (most basal leg segment) of the hind legs help guide and stabilize the female's ovipositor during oviposition.

Family Evaniidae

FAMILY	Evaniidae
OTHER NAMES	Evaniids, ensign wasps, hatchet wasps
DISTRIBUTION/WHERE FOUND	All continents but Antarctica (21 genera, 449 species)
SOLITARY OR SOCIAL?	Solitary
IDENTIFICATION	Small; abdomen flag-like or triangular, compressed laterally, attached high on thorax
SIMILAR WASPS	Usually mistaken for ants, but also small Sphecidae, Crabronidae
HABITAT	Outdoors where roaches occur; *Evania appendigaster* indoors
NEST	None
IMPORTANCE	Important parasitoid of egg capsules of domiciliary cockroaches

Ensign wasps are about as cute as wasps get. The abdomen waves or bobs up and down as they walk, so they seem to dance their way in search of the egg capsules of cockroaches. The life cycle of one species, *Evania appendigaster*, is chronicled on page 102. Since the single larva consumes multiple eggs of its cockroach host, evaniids are perhaps more properly defined as predators than external parasitoids. *Prosevania fuscipes* has also been widely distributed as a helpful control of indoor cockroaches.

Family Gasteruptiidae

FAMILY	Gasteruptiidae
OTHER NAMES	Gasteruptiids, wild carrot wasps
DISTRIBUTION/WHERE FOUND	All continents but Antarctica (6-9 genera, 496 species)
SOLITARY OR SOCIAL?	Solitary
IDENTIFICATION	Head on "neck"; flared and thickened tibia segment on hind leg ("leg warmers"); ovipositor may have white tip, long or short depending on species
SIMILAR WASPS	Aulacidae, Ichneumonidae, Braconidae
HABITAT	Variable, on umbelliferous flowers in fields and forest understory, and near nests of hosts.
NEST	None
IMPORTANCE	Sometimes a pest of "bee boards" housing leafcutter bees used to pollinate alfalfa and other crops

Wild carrot wasps might better be dubbed "giraffe wasps," due to their the elongated propleura, a segment at the front of the thorax that forms a "neck." Legs dangling when airborne, a female might be mistaken for a backwards-flying mosquito. She uses her ovipositor to lay eggs in the nests of solitary wasps and bees that utilize linear cavities and hollow twigs. The larval wasp may kill the host egg or larva and then eat the pollen or prey stored for it. Conversely, a gasteruptiid larva may wait patiently for the host to consume its cache of food, and *then* eat it.

Family Ceraphronidae

FAMILY	Ceraphronidae
OTHER NAMES	Ceraphronids
DISTRIBUTION/WHERE FOUND	All continents but Antarctica (15 genera, 304-360 species)
SOLITARY OR SOCIAL?	Solitary
IDENTIFICATION	Tiny (0.04–0.01 inches (1–3mm)), winged, wingless, or in between
SIMILAR WASPS	Other micro-Hymenoptera
HABITAT	Variable; many species living in soil
NEST	None
IMPORTANCE	Poorly known family, importance unknown

Known hosts for ceraphronid larvae include gall midge larvae, caterpillars, lacewing and scorpionfly larvae, mealybugs, and thrips. Some have been reared from the puparia of flies, and a few are hyperparasitoids of braconid wasp cocoons. Ceraphronids are internal parasites in all cases. Adults are perhaps found most often in soil, turning up in pitfall traps sunk in the ground.

Family Pelecinidae

FAMILY	Pelecinidae
OTHER NAMES	Pelecinids
DISTRIBUTION/WHERE FOUND	North America east of Rocky Mountains, South America (1 genus, 3 species)
SOLITARY OR SOCIAL?	Solitary
IDENTIFICATION	Large (0.8 inches (male)/2.8 inches (female) (20/70mm)); abdomen of female extraordinarily long, thin, bent into a semicircle; thickened and flared tibia segment on hind leg; males rare, resemble thread-waisted wasps
SIMILAR WASPS	Perhaps some Ichneumonidae, Sphecidae
HABITAT	Forest understory and forest edges
NEST	None
IMPORTANCE	No economic importance

These scarce wasps are among the most spectacular, the abdomen of the female being very long and slender. She plunges it into the soil to reach a grub of a May beetle (*Phyllophaga* spp., Scarabaeidae). Her larval offspring develop as internal parasitoids of the grub, probably overwintering as late first instar larvae (first larval stage after egg stage). Pupation occurs outside the host, but connected to it on its underside. Many populations of pelecinids appear to be parthenogenic, reproducing without benefit of male fertilization of the eggs. Consequently, males are rarely found.

Family Proctotrupidae

FAMILY	Proctotrupidae
OTHER NAMES	Proctotrupids
DISTRIBUTION/WHERE FOUND	All continents but Antarctica (~28 genera, 403 species)
SOLITARY OR SOCIAL?	Solitary, but larvae of some are gregarious parasitoids
IDENTIFICATION	Small (0.1–0.4 inches (3–10mm)); tip of abdomen downcurved in females, often joined to thorax with short, stout petiole (stalk)
SIMILAR WASPS	Some resemble Pompilidae, Crabronidae, alate (winged) male ants
HABITAT	Most common in wetlands and shaded habitats
NEST	None
IMPORTANCE	No economic importance

The larvae of most proctotrupids are internal parasitoids of fly larvae and beetle grubs living in the topsoil or decomposing wood. Several larvae may occupy a single host in some species. Pupation is outside the host, but the pupa remains attached to the belly of the host. The adult male wasps disperse farther than females, and females are most often encountered on or in the soil.

Family Diapriidae

FAMILY	Diapriidae
OTHER NAMES	Diapriids
DISTRIBUTION/WHERE FOUND	All continents but Antarctica (~190 genera, >2000 species)
SOLITARY OR SOCIAL?	Solitary
IDENTIFICATION	Tiny (0.04–0.3 inches (1–8mm) 0.08–0.15 inches average (2–4mm)); antennae arising from a "shelf" on the face; abdomen usually petiolate (stalked), ovipositor concealed
SIMILAR WASPS	More ant-like, but also Bethylidae, Braconidae, other microhymenoptera
HABITAT	Mostly damp, shaded habitats such as forest understory, wetlands, and in the soil
NEST	None
IMPORTANCE	No economic importance

Diapriids make up in diversity what they lack in size. The average person rarely sees them, though the wasps sometimes show up at blacklights at night. Most develop as endoparasitoids (internal parasitoids) of flies, the egg laid on the maggot and exiting as a new adult through the fly pupa, or completing the entire life cycle within the fly pupa. Other species are associated only with ants or termites, especially in the tropics.

Family Platygastridae

FAMILY	Platygastridae
OTHER NAMES	Platygastrids
DISTRIBUTION/WHERE FOUND	All continents but Antarctica (236 genera, 5385 species)
SOLITARY OR SOCIAL?	Solitary
IDENTIFICATION	Tiny (0.04–0.15 inches (1–4mm) usually 0.08 inches (2mm) or less; often sexually dimorphic, even polymorphic; winged, wingless, or in between
SIMILAR WASPS	Many other micro-Hymenoptera
HABITAT	Variable; many found in the soil, on flowers
NEST	None
IMPORTANCE	Occasionally used in agricultural pest control, with vast potential use in that regard

This family includes the subfamily Scelioninae, formerly a family of its own. It is estimated that the number of platygastrid species exceeds 7,000, with many known species awaiting description and many still undiscovered. Generalizations about life histories is almost useless, but generally platygastrids are at least initially egg parasitoids of insects and spiders. Members of the Platygastridae specialize mostly on gall midges. Some platygastrid larvae wait until the host reaches the pre-pupa or pupa stage before feeding as internal parasitoids.

Family Cynipidae

FAMILY	Cynipidae
OTHER NAMES	Cynipids, gall wasps
DISTRIBUTION/WHERE FOUND	All continents but Antarctica, most diverse and abundant in Europe and North America (74 genera, 1412 species)
SOLITARY OR SOCIAL?	Solitary
IDENTIFICATION	Small (0.04–0.3 inches (1–8mm)), compact; hunchback appearance with small head relative to thorax; sexually dimorphic, some species seasonally dimorphic, winged or wingless
SIMILAR WASPS	Other micro-Hymenopteran; ants
HABITAT	Forests, savannahs, and gardens, where host plants occur
NEST	Create galls on plants, especially oak trees
IMPORTANCE	Little economic importance in modern times, though cosmetic "damage" to hosts may occur; ecologically important in that galls sustain an array of other species

Gall wasps have been studied thoroughly and yet we learn more all the time. Mankind has made use of galls as food and folk medicine, and in the production of inks and dyes for centuries. The alternation of sexual and parthenogenic generations in some species confounds researchers to this day. In temperate climates, gall wasps are most conspicuous as adults in late autumn and early winter. Not all species form galls; some invade the galls of other gall wasps as uninvited guests (inquilines), but do no damage to their hosts.

Family Figitidae

FAMILY	Figitidae
OTHER NAMES	Figitids
DISTRIBUTION/WHERE FOUND	All continents but Antarctica (148 genera, 1571 species)
SOLITARY OR SOCIAL?	Solitary
IDENTIFICATION	No easily visible characteristics useful for field identification
SIMILAR WASPS	Nearly identical to many Cynipidae
HABITAT	Variable according to hosts; often around dung or rotting fruit
NEST	None, but may emerge from galls
IMPORTANCE	No economic importance

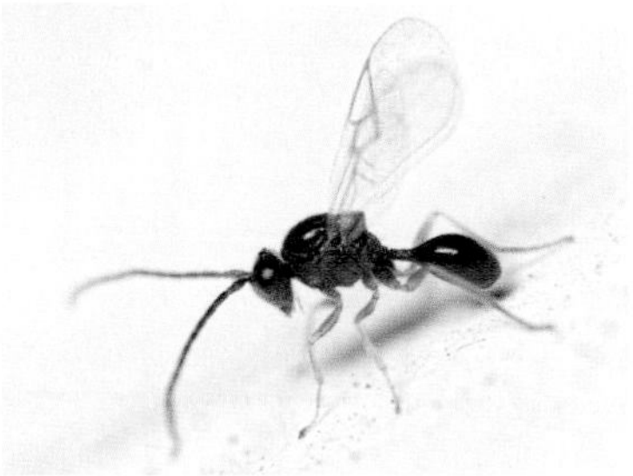

One subfamily of Figitidae, the Eucoilinae, was once a separate family. All are parasitoids of other insects, some specializing on gall wasps. Other known hosts include the larvae of lacewings (Chrysopidae and Hemerobiidae), the pupae of hover flies (Syrphidae), or the larvae and pupae of other flies. Figitid larvae are either endoparasitoids (internal parasitoids), or begin as endoparasitoids and finish as ectoparasitoids (external parasitoids).

Family Ibaliidae

FAMILY	Ibaliidae
OTHER NAMES	Ibaliids
DISTRIBUTION/WHERE FOUND	North America, Central America, western Europe, Japan, Australia, New Zealand (3 genera, 20 species)
SOLITARY OR SOCIAL?	Solitary
IDENTIFICATION	Large (1.18 inches (up to 30mm)); abdomen thin, compressed laterally
SIMILAR WASPS	Ichneumonidae, Braconidae
HABITAT	Mostly forests where wood-boring insect larvae are found
NEST	None
IMPORTANCE	No economic importance

Like many parasitoid wasps that are associated with wood-boring insect larvae, ibaliids are seldom seen except around freshly cut or windthrown timber. The hosts for these wasps are the grubs of horntail wood wasps (Siricidae) and cedar wood wasps (Anaxyelidae). The larvae are initially internal parasitoids of the host, but finish feeding externally. Ibaliids are by far the largest members of the Cynipoidea, which includes gall wasps and figitids.

Family Agaonidae

FAMILY	Agaonidae
OTHER NAMES	Agaonids, fig wasps
DISTRIBUTION/WHERE FOUND	All continents except Antarctica (20–30 genera, about 770 species)
SOLITARY OR SOCIAL?	Solitary, though gregarious inside figs
IDENTIFICATION	Tiny; extreme sexual dimorphism
SIMILAR WASPS	Other micro-Hymenoptera
HABITAT	Variable; dependent on location of host trees
NEST	None, but life cycle occurs inside a fig
IMPORTANCE	Obligate, exclusive pollinators of most figs

Fig wasps have extraordinarily complex life cycles, as generalized and illustrated on pages 90–91. The larval stage feeds within a single fig ovule, creating a gall. Males are larva-like and wingless, never leaving the confines of the fig. Ovules that do not receive a wasp egg develop into viable fig seeds. About half of fig species have male and female reproductive structures on separate trees. In these cases, galls are produced inside the male "fruits" only. Some wasps currently placed in this family are parasitoids of other fig wasps and will likely be reclassified into other wasp families. The parasitoids typically have long ovipositors and reach their hosts from the exterior of the fig. Still other agaonids carry out their life cycle without pollinating the plant.

Family Aphelinidae

FAMILY	Aphelinidae
OTHER NAMES	Aphelinids
DISTRIBUTION/WHERE FOUND	All continents except Antarctica (29-43 genera, 1078-1311 species)
SOLITARY OR SOCIAL?	Solitary
IDENTIFICATION	Tiny (0.08 inches or less (2mm)); no characteristics useful for field identification
SIMILAR WASPS	Other micro-Hymenopteran, especially Trichogrammatidae and Encyrtidae; Braconidae
HABITAT	Varies according to location of hosts; deployed in crop fields, orchards
NEST	None
IMPORTANCE	Reared extensively as biocontrol agents of pest insects

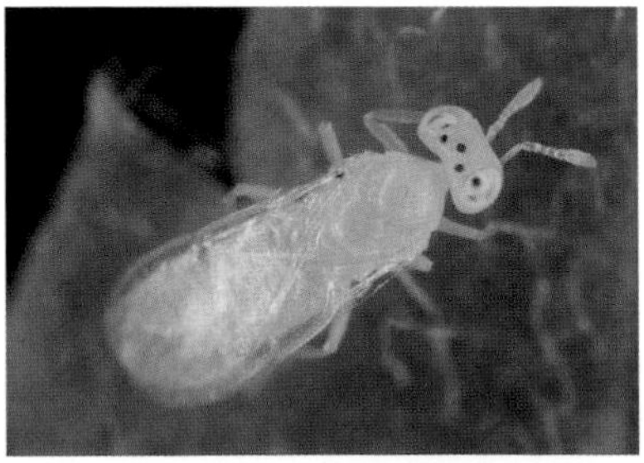

Aphelinids defy easy description, and the known life cycles are often bizarre. Many are vital agents in the control of scale insects, mealybugs, aphids, and whiteflies. Others target the eggs of moths, grasshoppers, and flies; or the larvae and pupae of flies. Still others attack larvae of other chalcidoid micro-Hymenopteran, or dryinid wasps, as hyperparasitoids. In some species, male and female aphelinids develop on different hosts. In a few cases, males are parasitoids of females of their own species. The larvae vary from internal to external parasitoids, or both.

Family Chalcididae

FAMILY	Chalcididae
OTHER NAMES	Chalcid wasps
DISTRIBUTION/WHERE FOUND	All continents except Antarctica (85-90 genera, ~1470 species)
SOLITARY OR SOCIAL?	Solitary, though many larvae may develop in a single host
IDENTIFICATION	Small (0.09–0.35 inches (2.5–9mm)), but large for icrohymenopteran; femur of hind leg greatly swollen, often toothed
SIMILAR WASPS	Leucospidae
HABITAT	Variable according to location of hosts
NEST	None
IMPORTANCE	No economic importance

Adult chalcids use those huge hind legs to jump, or battle female to female; or in the case of *Lasiochalcidia igiliensis*, to hold open the formidable jaws of an antlion larva while laying an egg on its throat. Most chalcids, as larvae, are internal parasitoids of the pupae of butterflies or moths, or the mature larvae of flies. A few species attack beetles, or other Hymenoptera. The adult mother wasp may lay her eggs on the eggs or larvae of the host, but her adult offspring emerge from the host's pupa stage.

Family Encyrtidae

FAMILY	Encyrtidae
OTHER NAMES	Encyrtids
DISTRIBUTION/WHERE FOUND	All continents except Antarctica, most oceanic islands (493-506 genera, ~4100 species)
SOLITARY OR SOCIAL?	Solitary
IDENTIFICATION	Small (0.04–0.15 inches (1–4mm)); long spur at tip of tibia on middle leg; sometimes colorful, with banded wings
SIMILAR WASPS	Other micro-Hymenonptera, ants
HABITAT	Varies with location of hosts
NEST	None
IMPORTANCE	Significant agents of control of pest scale insects, some species introduced to foreign countries for that purpose.

At least half of the encyrtids are internal parasitoids of immature scale insects or mealybugs, sometimes adult scales. Many others are gregarious internal parasitoids of caterpillars. These species may develop from a single egg (polyembryony) laid by the mother wasp. Still other encyrtid species are parasitoids of other insects, or mites, ticks, or spiders. A few are hyperparasitoids of other micro-Hymenopteran, Braconidae, Dryinidae that are living inside other insects. Encyrtids usually pupate inside the host, in some cases integrating a membraneous envelope into the tracheal system of the still-living host.

Family Eucharitidae

FAMILY	Eucharitidae
OTHER NAMES	Eucharitids; ant parasitoids
DISTRIBUTION/WHERE FOUND	All continents except Antarctica; most diverse in tropics and subtropics (57–68 genera, about 430 species)
SOLITARY OR SOCIAL?	Solitary
IDENTIFICATION	Small (0.08–0.2 inches (2–5.4mm)); antennae often pectinate (comb-like) or flabellate (fan-like or plate-like); thorax often adorned with prongs or spines; abdomen often unusually small, attached to thorax by long petiole (stalk)
SIMILAR WASPS	Some other micro-Hymenoptera
HABITAT	Varies with location of hosts: ant colonies
NEST	None
IMPORTANCE	No economic importance, though potential for use in control of invasive ant species

Eucharitids are, as far as we know, exclusively ant parasitoids. The larvae feed internally or externally on ant larvae or pupae, depending on the species. Larvae of some eucharitids can jump.

Family Eulophidae

FAMILY	Eulophidae
OTHER NAMES	Eulophids
DISTRIBUTION/WHERE FOUND	All continents except Antarctica, most oceanic islands (~330 genera, 4969–5197 species)
SOLITARY OR SOCIAL?	Solitary, though larvae of some are gregarious parasitoids
IDENTIFICATION	No characteristics that are useful for identification in the field
SIMILAR WASPS	Other microhymenoptera
HABITAT	Varies according to location of the host
NEST	None
IMPORTANCE	Some species are utilized as biocontrol agents of other pest insects

The bulk of eulophid species are generalist parasitoids of insect larvae that are concealed, especially those that are leaf miners, living between the layers of a leaf. These hosts include larvae of flies, moths, and beetles. Some eulophid larvae are predatory, feeding on spider eggs in an egg sac, for example, or mites inside a gall. In short, eulophids may exhibit the widest degree of parasitoid lifestyles, and host ranges, of all microhymenopterans. A small number are plant feeders or gall-makers. *Melittobia* eulophids are detailed on page 194.

Family Eupelmidae

FAMILY	Eupelmidae
OTHER NAMES	Eupelmids
DISTRIBUTION/WHERE FOUND	All continents except Antarctica, most oceanic islands (~50 genera, >930 species)
SOLITARY OR SOCIAL?	Solitary, though some are gregarious as larvae
IDENTIFICATION	Small (<0.4 inches (<10mm)); body often slender, elongate; some species metallic; wings of some species greatly reduced
SIMILAR WASPS	Other micro-Hymenoptera, especially Encyrtidae, Pteromalidae
HABITAT	Varies according to host; most visible on dead, standing trees and logs
NEST	None
IMPORTANCE	No economic importance

Members of the subfamily Eupelminae have drastic modifications to the middle section of the thorax to allow prodigious jumping ability. In death these large muscles contract, resulting in a contorted, backward arched posture with head nearly meeting abdomen. Metapelma are encountered on dead, standing trees or logs where they search for wood-boring beetle hosts. The diversity of eupelmid hosts, and variety of life cycles, rivals that of the Eulophidae. Some are plastic enough to be hyperparasitoids of other wasp larvae inside a primary host insect.

Family Eurytomidae

FAMILY	Eurytomidae
OTHER NAMES	Eurytomids
DISTRIBUTION/WHERE FOUND	All continents except Antarctica (80–97 genera, ~1500 species)
SOLITARY OR SOCIAL?	Solitary, though a handful are gregarious as larval parasitoids
IDENTIFICATION	Small (0.05–0.2 inches (1.4–6mm)); no characteristics suitable for identification in the field
SIMILAR WASPS	Other micro-Hymenoptera
HABITAT	Variable, according to where hosts are found
NEST	None
IMPORTANCE	Some "seed chalcids" are important crop pests

Eurytomids are so diverse in physical appearance as to be confused with every other family of micro-Hymenopteran. They exhibit equally diverse lifestyles. In the larval stage, most are concealed. Seed-feeders are largely restricted to legumes, and plants with umbelliferous flowers. Stem borers afflict grasses and some herbs, some species forming stem galls that stunt the growth of the flower head. A number of eurytomids are, as larvae, external parasitoids of gall wasp larvae and the immature stages of other gall-formers. A few of these feed on the gall itself before, during, or after feeding on the host insect. A small number of eurytomids are facultative hyperparasitoids that can, if they choose, become parasitic on other ectoparasitoids.

Family Leucospidae

FAMILY	Leucospidae
OTHER NAMES	Leucospids
DISTRIBUTION/WHERE FOUND	All continents except Antarctica (4 genera, 134–151 species)
SOLITARY OR SOCIAL?	Solitary
IDENTIFICATION	Large (0.15–0.66 inches (4–17mm)); colorful, usually black and yellow or red; wings folded lengthwise at rest; ovipositor of female curved over top of abdomen; femur of hind leg greatly swollen, toothed
SIMILAR WASPS	Chalcididae, Vespidae
HABITAT	Variable, where hosts are found; conspicuous flower visitors
NEST	None
IMPORTANCE	Occasional pest of "bee boards" housing leafcutter bees used to pollinate alfalfa and other crops

Leucospids are among the largest microhymenopterans, but easily mistaken for potter and mason wasps. Unique articulation of the abdomen and configuration of the ovipositor allowsS the female wasp to penetrate dense substrates to reach immature solitary bees, and some stinging parasitoid wasps, which nest in pre-existing cavities in wood or twigs. The larval leucospid first eliminates competing parasitoids before settling to feed on the body fluids of the host as an external parasitoid.

Family Mymaridae

FAMILY	Mymaridae
OTHER NAMES	Mymarids, fairyflies
DISTRIBUTION/WHERE FOUND	All continents except Antarctica, most oceanic islands (96–117 genera, ~1450 species)
SOLITARY OR SOCIAL?	Solitary, though a few are gregarious as larvae
IDENTIFICATION	Minute (most <0.04 inches (<1mm)); wing veins absent; wings often fringed with long setae (hairs); hind wing may be reduced to a "paddle" or filament; some species are wingless
SIMILAR WASPS	Other nearly micro-Hymenoptera
HABITAT	Varies according to host, including aquatic habitats
NEST	None
IMPORTANCE	Some species have been widely introduced for control of agricultural pests

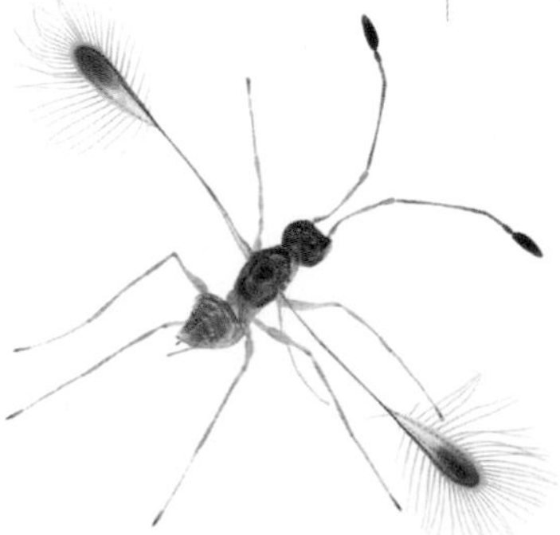

Fairyflies, it is said, could fly through the eye of a needle, so diminutive are most of them. All species are internal parasitoids of the eggs of other insects. Most mymarids target eggs by their habitat and location, rather than a particular host species. Eggs concealed partly or wholly in plant tissues, or in the soil, are typically exploited. Most species overwinter as mature larvae inside the host egg. Pupation also occurs within the egg.

Family Perilampidae

FAMILY	Perilampidae
OTHER NAMES	Perilampids
DISTRIBUTION/WHERE FOUND	All continents except Antarctica (17 genera, 284–292 species)
SOLITARY OR SOCIAL?	Solitary
IDENTIFICATION	Small (0.05–0.2 inches (1.3–5.5mm)), compact; abdomen pyramidal in shape; thorax often densely sculptured; many species metallic green or blue
SIMILAR WASPS	Often mistaken for Chrysididae cuckoo wasps, sometimes metallic bees
HABITAT	Varies according to host; conspicuous on flowers, around aphid colonies
NEST	None
IMPORTANCE	No economic importance

Most perilampids are hyperparasitoids: parasitoids of parasitoids within a primary host. Primary hosts include caterpillars, sawfly larvae, nymphs of grasshoppers, lacewings, and other Hymenoptera. The intended targets are larvae of tachinid flies (Tachinidae), ichneumon wasps, or braconid wasps. A few perilampids can make use of the primary host, others parasitoids of certain wood-boring beetles. The female wasps lay a multitude of eggs in or on plant tissue, and the active larvae that hatch (planidia) must latch onto another insect in the hope it is a host or a vehicle to get to a host. If successful, it enters the host in search of another parasitoid.

Family Pteromalidae

FAMILY	Pteromalidae
OTHER NAMES	Pteromalids
DISTRIBUTION/WHERE FOUND	All continents except Antarctica, most oceanic islands (~630 genera, about 3600 species)
SOLITARY OR SOCIAL?	Solitary, but some are gregarious as larvae
IDENTIFICATION	Small to very large (0.04–1.9 inches (1–48mm)); often metallic; diversity of body shapes; winged or wingless
SIMILAR WASPS	Many other micro-Hymenoptera
HABITAT	Varies according to host
NEST	None
IMPORTANCE	Often employed as biological control agents for agricultural pests; *Pteromalus puparum* can be a pest of butterfly farms

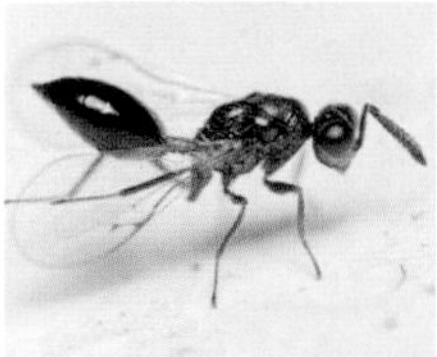

The Pteromalidae is largely a catch-all family of misfit wasps that cannot be easily classified elsewhere. Their variety in physical appearance and disparate lifestyles reflects this. Most are idiobionts, meaning they arrest further development of the host, or attack it during the prepupa or pupa stage. Females of some lay eggs in the host larva, her adult offspring exiting from the host pupa. Many species attack gall-making insects, or stem borers, leaf miners, or wood borers. *Pteromalus puparum* attacks the chrysalids of butterflies. A surprising number of pteromalid larvae are predators, feasting on insect eggs. There is scarcely any insect that cannot be a pteromalid host.

Family Torymidae

FAMILY	Torymidae
OTHER NAMES	Torymids
DISTRIBUTION/WHERE FOUND	All continents except Antarctica, many oceanic islands (68–82 genera, >900 species)
SOLITARY OR SOCIAL?	Solitary
IDENTIFICATION	Small (0.4–0.3 inches (1–7.5mm), excluding ovipositor); many are exquisitely metallic; ovipositor of female often long, sweeping upward
SIMILAR WASPS	Other microhymenoptera
HABITAT	Variable; many species associated with galls
NEST	None, but may emerge from galls
IMPORTANCE	Some "seed chalcids" are important crop and orchard pests

The torymids most often seen are those in the subfamily Toryminae, which are associated with galls. Most are primary external parasitoids of prepupa or pupa of those gall-formers. A few are inquilines, feeding on gall tissue without harming the host, while some eat both the occupant and the gall. Some torymids are parasitoids of cocooned prepupae of square-headed wasps (Crabronidae) nesting in hollow twigs. Torymids have even more fascinating life histories, too numerous to detail here.

Family Trichogrammatidae

FAMILY	Trichogrammatidae
OTHER NAMES	Trichogrammatids
DISTRIBUTION/WHERE FOUND	All continents except Antarctica, many oceanic islands (98 genera, >880 species)
SOLITARY OR SOCIAL?	Solitary, though some are gregarious as larvae
IDENTIFICATION	Tiny (0.01–0.04 inches (0.3–1.2mm))
SIMILAR WASPS	Mymaridae
HABITAT	Varies according to host, including aquatic habitats
NEST	None
IMPORTANCE	Highly effective agents of biocontrol of agricultural and garden pests; commercially reared and available in retail outlets

Trichogrammatids are parasitoids of the eggs of other insects, especially those of butterflies and moths, true bugs, beetles, thrips, flies, lacewings and their kin, and other wasps. Some attack aquatic hosts such as dragonflies or diving beetles. Females of some trichogrammatids ride the adult female host until she lays eggs (phoretic oviposition). Trichogrammatids identify appropriate host eggs largely by touch and smell. These tiny wasps are mostly at the mercy of the wind. One species, *Megaphragma mymaripenne*, is so small that it has as few as 7,400 neurons, which function without nuclei.

Family Braconidae

FAMILY	Braconidae
OTHER NAMES	Braconids
DISTRIBUTION/WHERE FOUND	All continents except Antarctica, most oceanic islands (1057 genera, 17600–19205 species)
SOLITARY OR SOCIAL?	Solitary, though some are gregarious as larvae
IDENTIFICATION	Mostly small (0.07–0.6 inches (2–15mm) excluding ovipositor); wing venation distinctive when visible; flagellum of antenna divided into many short flagellomeres
SIMILAR WASPS	Ichneumonidae
HABITAT	Most habitats; look for them on flowers, foliage, and logs
NEST	None
IMPORTANCE	Several are used as biological control agents of pests like the codling moth and imported cabbageworm butterfly

The mechanisms of parasitism in Braconidae are incredibly complex. Most are endoparasitoid koinobionts: internal parasitoids that allow the host animal to continue growing. Many are egg-larva parasitoids, ovipositing in the host egg, but developing in the host larva. A large percentage of others are ectoparasitoid idiobionts: the female wasp stings the host into temporary paralysis, lays an egg, and her offspring begins feeding on the host immediately. Common hosts for braconids include caterpillars, beetles, and flies, as well as aphids and other insects with simple metamorphosis. Some braconids deploy a polydna bracovirus to control the behavior of their host (page 95). *Dinocampus coccinellae* ravages the inside of a lady beetle host before enslaving it as a "zombie" guardian of the wasp's cocoon.

Family Ichneumonidae

FAMILY	Ichneumonidae
OTHER NAMES	Ichneumons, Darwin wasps
DISTRIBUTION/WHERE FOUND	All continents except Antarctica, most oceanic islands (1575 genera, ~24025 species)
SOLITARY OR SOCIAL?	Solitary, though more than one larva may feed on a host
IDENTIFICATION	Small to very large (0.11–1.65 inches (3–42mm) excluding ovipositor); wing venation distinctive: often a "horse head" cell in upper center of forewing; flagellum of antenna divided into many short flagellomeres
SIMILAR WASPS	Braconidae, Aulacidae, Gasteruptiidae, Sphecidae, Pompilidae
HABITAT	Most habitats, especially forests and forest edges
NEST	None
IMPORTANCE	Natural biological control agents of many pest insects

This largest family in the order Hymenoptera features a wide range of hosts and parasitoid lifestyles. Most commonly, ichneumons are internal parasitoids of caterpillars and sawfly larvae, but immature stages of many other insects are exploited, including beetles, flies, other wasps, snakeflies, and caddisflies. A surprising number are external parasitoids of spiders. Some, such as *Reclinervellus nielseni*, program the spider to spin a special web for the wasp larva to spin a cocoon upon, suspended and protected from its own enemies. A few are predators within the egg sacs of spiders or pseudoscorpions. At a 2019 convention of ichneumon wasp specialists, the scientists assigned "Darwin wasps" as a new common name for the family.

Family Bethylidae

FAMILY	Bethylidae
OTHER NAMES	Flat wasps, bethylids
DISTRIBUTION/WHERE FOUND	All continents except Antarctica, most oceanic islands (84 genera, 2340 species)
SOLITARY OR SOCIAL?	Solitary, females of some species guard their young
IDENTIFICATION	Small (0.03–0.78 inches (1–20mm)); many are wingless and ant-like; black or brown
SIMILAR WASPS	More likely to be mistaken for ants
HABITAT	Variable, including grain storage facilities, houses
NEST	None, though some hide hosts in pre-existing cavities
IMPORTANCE	Some are important agents of biocontrol for stored product pests like moths and beetles

Flat wasps are common, but not often encountered. The known hosts for these stinging wasps are apparently limited to certain caterpillars and beetle larvae, especially those infesting stored grains. A few seek hosts that are sheltered in some fashion, such as wood-borers, soil-dwellers, case-bearing leaf beetle larvae, and leaf-rolling caterpillars. Bethylids sometimes move the incapacitated host into a crevice or other nook. Females of a few genera guard their small clutch of eggs and the larvae that hatch from them.

Family Chrysididae

FAMILY	Chrysididae
OTHER NAMES	Chrysidids, cuckoo wasps, gold wasps, ruby wasps, jewel wasps
DISTRIBUTION/WHERE FOUND	All continents except Antarctica, many oceanic islands (81 genera, ~2500 species)
SOLITARY OR SOCIAL?	Solitary
IDENTIFICATION	Small (0.11–0.7 inches (3–18mm) mostly <0.2 inches (<6mm)); most are brilliant metallic green, blue, copper, bronze, and/or red; body heavily sculptured, pitted; a few are wingless
SIMILAR WASPS	Perilampidae, and many metallic bees
HABITAT	Most habitats, including rural, suburban and urban areas
NEST	None, but seen around nests of other wasps
IMPORTANCE	No economic importance

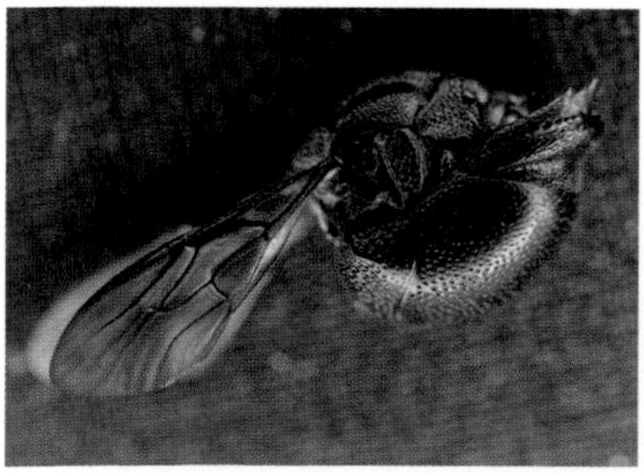

Cuckoo wasps do not sting. The females of the most familiar varieties have a telescoping ovipositor formed by several terminal abdominal segments. This anatomical feature is used to lay an egg in the nest of another solitary wasp or bee. The chrysidid larva is either a cleptoparasite that eats the provisions provided for the host, or is a parasitoid on the host itself. The subfamilies Amiseginae and Loboscelidiinae are non-metallic, not heavily armored, and are parasitoids of walking stick eggs. Cleptinae are parasitoids of prepupal sawfly larvae (Diprionidae and Tenthredinidae).

Family Sphecidae

FAMILY	Sphecidae
OTHER NAMES	Sphecids, thread-waisted wasps, mud daubers, digger wasps
DISTRIBUTION/WHERE FOUND	All continents except Antarctica, most oceanic islands (19 genera, 724 species)
SOLITARY OR SOCIAL?	Solitary, though females may nest in close proximity
IDENTIFICATION	Large (0.4–1.18 inches (10–30mm)); long, narrow petiole connecting gaster to thorax; tibia of middle leg with two spurs at the tip
SIMILAR WASPS	Some Ichneumonidae, Aulacidae, Gasteruptiidae, Stephanidae, Pelecinidae, Vespidae, and Crabronidae
HABITAT	Found in most habitats, including urban and suburban
NEST	Burrows in the soil, pre-existing cavities, free-standing mud nests
IMPORTANCE	Major predators of pest caterpillars, grasshoppers; pollinators

Sphecids are solitary, stinging parasitoids. Each female digs a burrow for her nest, uses a pre-existing cavity as a nest, or constructs free-standing mud nests of one or more individual cells. Prey varies according to genus and species. Spiders, cockroaches, crickets, katydids, grasshoppers, and caterpillars are the primary hosts. Adult sphecids are frequent visitors to flowers, and are fond of honeydew secreted by aphids and related insects. Some species congregate in loose or dense clusters to spend the night, hanging from vegetation or concentrating in nooks and sheltered locations.

Family Crabronidae

FAMILY	Crabronidae
OTHER NAMES	Crabonids, sand wasps, digger wasps, square-headed wasps
DISTRIBUTION/WHERE FOUND	All continents except Antarctica, most oceanic islands (242 genera, ~8700 species)
SOLITARY OR SOCIAL?	Solitary, a few primitively social; females of some species nest in large aggregations
IDENTIFICATION	Small to very large (0.06–1.77 inches (1.5–45mm)); highly variable body shape and coloration defy easy identification
SIMILAR WASPS	Vespidae, Sphecidae, Tenthredinidae, Pompilidae
HABITAT	Most habitats
NEST	Burrows in soil, pre-existing cavities, free-standing mud nests; several are cleptoparasites or parasitoids in nests of other wasps
IMPORTANCE	Minor importance in biocontrol of pests, monitoring invasive species (*Cerceris fumipennis* for emerald ash borer)

Crabronids are the most diverse of the stinging parasitoid wasps, and classification of this large family has been problematic. They were once lumped under Sphecidae, while some authorities elevate crabronid subfamilies to family-level. Cicada killers, beewolves, horse guards, and the organ pipe mud dauber are familiar examples of the varied life histories of crabronids. All are stinging parasitoids and many species display elaborate behavior in courtship, host-seeking, and nesting.

Family Mutillidae

FAMILY	Mutillidae
OTHER NAMES	Mutillids, velvet ants
DISTRIBUTION/WHERE FOUND	All continents except Antarctica, many oceanic islands (210 genera, ~4300 species)
SOLITARY OR SOCIAL?	Solitary
IDENTIFICATION	Small to medium (0.11–0.98 inches (3–25mm)); sexually dimorphic: females wingless, males winged (usually); often densely hairy; diurnal species brightly colored
SIMILAR WASPS	Chyphotidae, Bradynobaennidae, Myrmosidae, some Tiphiidae, Thynnidae
HABITAT	Most habitats, but especially conspicuous in deserts, dunes and other arid situations
NEST	None, but parasitoids in the nests of other wasps and bees
IMPORTANCE	No economic importance; sting of female is excruciating

The fuzzy appearance of many velvet ants belies the powerful sting of the female wasp. They are the most conspicuous of wingless wasps, though many desert species are crepuscular (active at dawn and/or dusk) or nocturnal. Males are often larger than females, and fly low over the ground in search of females. Larvae of mutillids are mostly ectoparasitoids (external parasitoids) of the larvae or pupae of other solitary wasps or bees, though other hosts are recorded. See page 82 for more about velvet ants.

Family Tiphiidae

FAMILY	Tiphiidae
OTHER NAMES	Flower wasps, tiphiids
DISTRIBUTION/WHERE FOUND	All continents except Antarctica, many oceanic islands (24 genera, 350 species)
SOLITARY OR SOCIAL?	Solitary
IDENTIFICATION	Small (0.98 inches (25mm) and less); stout-bodied, often hairy; legs spinose
SIMILAR WASPS	Thynnidae, some Scoliidae, Mutillidae, Myrmosidae, Chyphotidae, Bradynobaenidae
HABITAT	Most habitats
NEST	None
IMPORTANCE	Important biocontrol agent for Japanese beetle

Tiphiids can be abundant wasps on flowers, especially in arid habitats, in spring and late summer. As larvae they are external parasitoids on grubs of scarab beetles. Females of some species are wingless. She digs to find a host, stings it into temporary paralysis, and then lays an egg on it. The spring tiphia, *Tiphia vernalis*, was introduced to the U.S.A. in 1925 from Korea and China in the hope of controlling the Japanese beetle.

Family Thynnidae

FAMILY	Thynnidae
OTHER NAMES	Flower wasps, thynnids
DISTRIBUTION/WHERE FOUND	Australia, east Asia, North and South America (63–75 genera, >222 species)
SOLITARY OR SOCIAL?	Solitary, but males may sleep in large, loose clusters
IDENTIFICATION	Small (<1.18 inches (<30mm)); strongly sexually dimorphic; females of many are wingless
SIMILAR WASPS	Tiphiidae, Sphecidae, Crabronidae, Vespidae
HABITAT	Most habitats, most conspicuous in open areas
NEST	None
IMPORTANCE	No economic importance

Thynnids were split off from the family Tiphiidae in 2008. They are most diverse in Australia, but members of the genus *Myzinum* are common in North America. Most are, as larvae, ectoparasitoids on subterranean beetle larvae, especially scarabs. *Methocha* species are parasitoids of the larvae of tiger beetles, requiring great fearlessness and agility on the part of the wingless female wasp. Males of some thynnids practice phoretic copulation.

Family Scoliidae

FAMILY	Scoliidae
OTHER NAMES	Mammoth wasps, scoliids
DISTRIBUTION/WHERE FOUND	All continents except Antarctica, most oceanic islands (143 genera, 560 species)
SOLITARY OR SOCIAL?	Solitary, though males may sleep in loose groups
IDENTIFICATION	Small to gigantic (<0.04–>0.15 inches (<1cm–>4cm)); Wings iridescent, with numerous fine creases on the outer third of the forewing; legs very spiny; body robust, often hairy; often colorful; sexually dimorphic
SIMILAR WASPS	Tiphiidae
HABITAT	Most habitats
NEST	None
IMPORTANCE	No economic importance, though helpful allies in controlling sod-infesting beetle groups

Scoliids are most diverse and conspicuous in the tropics, where some reach mammoth proportions. The sexes may look very different (sexual dimorphism), males often with a pseudosting at the posterior of the abdomen, and much longer antennae than females. Larvae of all species are ectoparasitoids (external parasitoids) of scarab beetle larvae or, rarely, weevil larvae.

Family Pompilidae

FAMILY	Pompilidae
OTHER NAMES	Spider wasps, tarantula hawks, pompilids
DISTRIBUTION/WHERE FOUND	All continents except Antarctica, most oceanic islands (125 genera, 4855 species)
SOLITARY OR SOCIAL?	Solitary
IDENTIFICATION	Small to gargantuan (<0.4–1.96 inches (<10mm–50mm)); long-legged, with two long spines at tip of hind tibia segment
SIMILAR WASPS	Ichneumonidae, Braconidae, Sphecidae, Crabronidae, Tenthredinidae
HABITAT	Most habitats
NEST	Burrow in the soil, pre-existing cavity, free-standing mud cells
IMPORTANCE	No economic importance

Spider wasps spend as much time running as they do flying. Females are in a constant, feverish search for spider prey and with flicking wings and bobbing antennae, they appear nervous; if you were hunting spiders, you would be, too. Typically, the wasp stings a single spider into permanent paralysis, then totes it to a nest she has already prepared. This may be a burrow, a pre-existing cavity, or in the case of the Auplopini tribe, a mud cell the wasp builds. She lays a single egg, seals the nest, and repeats the process. Pompilids in the genus *Evagetes*, and subfamily Ceropalinae, do not hunt, but are kleptoparasitoids in the nests of other pompilids.

Family Sapygidae

FAMILY	Sapygidae
OTHER NAMES	Sapygids
DISTRIBUTION/WHERE FOUND	North America, South America, Europe, and parts of Africa and Asia (12 genera, 66 species)
SOLITARY OR SOCIAL?	Solitary
IDENTIFICATION	Small (~0.19–0.75 inches (~5–19mm)); slender bodied, black with white or yellow bands
SIMILAR WASPS	Thynnidae, Crabronidae, Vespidae
HABITAT	Most habitats where host bees and wasps occur
NEST	None, but common around nests of solitary bees
IMPORTANCE	Occasional pest of "bee boards" housing leafcutter bees used to pollinate alfalfa and other crops

The larvae of all known sapygids are external parasitoids or kleptoparasitoids in the nests of solitary bees and mason wasps that nest in pre-existing, linear cavities, or sometimes in burrows in the ground. The handsome adult wasps perch where there are a number of active nests, patiently waiting for an opportunity to sneak in and lay their eggs. Sapygids will also visit flowers for nectar.

Family Vespidae

FAMILY	Vespidae
OTHER NAMES	Mason wasps, potter wasps, pollen wasps, hover wasps, paper wasps, yellowjackets, hornets, vespids
DISTRIBUTION/WHERE FOUND	The entire planet, except Antarctica (268 genera, 4932 species)
SOLITARY OR SOCIAL?	Solitary, primitively social, and truly social
IDENTIFICATION	Small to very large; face usually distinctly triangular; wings folded lengthwise at rest (except Masarinae); often colorful
SIMILAR WASPS	Sphecidae, Crabronidae, Tenthredinidae
HABITAT	Most habitats, including rural, suburban, and urban areas
NEST	Solitary species in burrows in soil, pre-existing cavities, free-standing mud nests; social species in mud or paper nests
IMPORTANCE	Notorious pests of urban areas, fruit orchards, beehives; scavenging species compete for our food and beverages

These are the wasps we think of when we hear the word "wasp." The family encompasses the full range of behaviors of stinging parasitoids, plus the complex social behavior of predators that bring food to their larvae as needed. It is a wonder that any caterpillar or fly escapes these hunters. It is equally amazing that we usually escape being stung, though that is a testament to the tolerance of the wasps more so than our tolerance for them.

Wasp Watching

Should you be so inclined to pursue the study of wasps in your leisure time, here are some recommendations for doing so safely and productively.

- Determine whether you are potentially allergic to insect venoms and take appropriate precautions.
- Outfit yourself with appropriate equipment such as close-focusing binoculars, pocket notebook and pen, insect net, plastic vials or similar containers, field guides, camera and/or smart phone, additional GPS device. You may deem that a telescope is safest for observing yellowjacket and hornet nests.
- Record observations with still images or video. Always note the date and the location of your observations as precisely as possible (country, province or state, township; latitude and longitude is even better). Share your findings via iNaturalist or other media.
- Capture and retention of specimens is optional, but done in the correct manner can be valuable scientifically. It is often the only means of identifying the species you are observing. Consult the advice of a local entomologist.
- Some wasps are nocturnal. Turn on the porch light to attract them, or invest in blacklighting equipment. Enjoy seeing the non-wasps drawn there, too.
- Build a "bee condo" or "bee block" and deploy it appropriately. Watch for mason wasps (Vespidae: Eumeninae), some spider wasps (Pompilidae), thread-waisted wasps (Sphecidae), and cavity-nesting square-headed wasps (Crabronidae) using the artificial housing. Cuckoo wasps (Chrysididae), sapygids (Sapygidae), wild carrot wasps (Gasteruptiidae), leucospids (Leucospidae), and other parasitoid wasps may attempt to infiltrate the nests of the solitary bees and wasps nesting there.
- The activities of many paper wasps can be observed at close range if you recognize and respect their body language. Back off if a wasp stands on tiptoe and flares her wings. Make use of those binoculars, too.
- Solitary wasps that fly from underfoot while you are hiking a sandy trail are probably in the act of nesting. Stop, back up, and see if they resume their business. Find a comfortable place to sit and watch them dig.
- Watch for male wasps sitting on prominent objects, on stones along trails, or on the ground. They will tend to return to the very same spot.
- Look on the broad leaves of sunlit plants for wasps pausing to groom themselves. Stake out the edge of a pond or puddle and watch for wasps coming to drink.
- Do not neglect galls, especially those on oak trees. Collect ones that show no emergence holes, confine them in a container, and record what emerges.
- Have fun and enjoy finding fascination in the natural world as a whole—wasps do not exist in a vacuum.

Watching You
A paper wasp peers from a hiding place. Wasps are far more alert to our presence than we are to theirs. A little human curiosity can lead to a lifetime of wonder and discovery.

Glossary

abdomen Third posterior division of the insect body. In wasps, the apparent abdomen plus the propodeum. *See* metasoma, gaster. Internally, contains vital organs of digestion, excretion, and reproduction.

Aculeata Informal subset of Apocrita comprised of all stinging wasps, bees, and ants. Not considered a formal taxonomic category.

aggregation Gathering of normally solitary individuals that falsely suggests the species is social.

Apocrita Suborder of Hymenoptera including all wasps, bees, and ants, exclusive of sawflies, horntails, and their kin (Symphyta).

aposematism Self-defense through bright color combinations advertising distastefulness, toxicity, or the ability to sting.

Batesian mimicry Resemblance of a harmless, "tasty" species to a toxic or well-defended species in physical appearance, behavior, or both.

beewolf Any solitary wasp belonging to the genus *Philanthus*.

biocontrol Employment of insects, bacteria, viruses and other pathogens by man in the control of pest insects.

bracovirus A virus peculiar to some wasps in the Braconidae that exerts control over the host's immune and/or nervous system for benefit of the parasitoid's offspring. A polydna virus.

brood Eggs, larvae, and pupae of a wasp, especially the collective brood in a nest of social wasps.

calyx Expansion of the oviduct into which the ovaries open. Where bracoviruses replicate inside certain braconid wasps.

cell Individual chamber for a wasp in the egg, larva, or pupa stage.

Chalcidoidea Superfamily that contains most families of small, non-stinging parasitoid wasps.

cladogram A diagram illustrating suspected relationships between separate taxonomic units of organisms, informed by sequencing of DNA or RNA, with or without the complement of observable physical characteristics. Not equivalent to an evolutionary tree.

colony Population unit of related individuals in a nest of social insects.

crop Stomach-like internal organ for storage of nectar and other liquid in an adult wasp. Contents of the crop are usually regurgitated to feed larval offspring, and/or nestmates in social wasps.

dominance hierarchy Ranking of female social wasps in a colony. The most dominant is usually the sole reproductive individual. *See* gyne.

ectoparasitoid A parasitoid that feeds on a host externally. An exposed parasitoid.

endoparasitoid A parasitoid that feeds on a host internally. A concealed parasitoid.

eusocial Truly social lifestyle that involves division of labor (including reproduction), and an overlap of generations inside one colony.

fossorial Lifestyle and behavior of an animal that digs burrows in soil.

gaster Abdomen of a wasp, exclusive of the propodeum. *See* metasoma.

genus Unit of taxonomy that represents a collection of subgenera, and species. A subset of tribe, subfamily, and family. Plural: genera.

gyne Reproductive female member of a colony of social insects. Queen. Gynes may or may not be physically different from other females in a colony, but they are at the top of the dominance hierarchy.

haplodiploidy Sex determination mechanism in all Hymenoptera whereby females have two sets of chromosomes (diploid), from fertilized eggs, but males have one set (haploid), from unfertilized eggs.

holometabolous Insects undergoing complete metamorphosis from egg to larva to pupa to adult.

hornet Any eusocial wasp that is a member of the genus *Vespa*.

host Any organism that supports a parasite or a parasitoid.

Hymenoptera Taxonomic order to which all wasp families belong, together with bees and ants.

hyperparasitoid A parasitoid of a parasitoid. A secondary parasitoid.

idiobiont Parasitoid that arrests further development of the host at the time it oviposits by stinging it into permanent paralysis.

inquiline Uninvited "guest" species in the home of a host organism, usually causing no harm to the host.

insectary A physical facility devoted to the breeding of beneficial insects for use in pest control.

instar Interval between molts of the insect larva. The first instar is what emerges from the egg, the second instar occurs after the first molt, and so forth.

juvenile hormone Principal hormone governing growth and development of insects. Abbreviated JH. The presence of JH manifests as a continued immature state of the insect. Reduction or absence of JH results in advance to the next stage in the life cycle, or adulthood.

kleptoparasitoid A parasitoid that develops on food stored for its host by the host's parent, resulting in the death of the host.

koinobiont Parasitoid that permits continued development of the host after oviposition into the host (sometimes after stinging the host into temporary paralysis).

lek A location where male animals gather to display for females.

mass provisioning Caching of a quantity of food

by a female solitary stinging parasitoid wasp for her larval offspring.

mesosoma. In wasps, the traditional thorax, plus the propodeum (first segment of the abdomen).

metasoma Abdomen of a wasp, exclusive of the propodeum. *See* gaster.

mimic Organism passing itself off as another, unrelated organism (model) through physical resemblance and/or behavior, for purposes of self-defense.

model Organism that is distasteful, toxic, or well-defended that serves as a template for the evolution of mimics.

mud dauber Solitary stinging parasitoids in the families Sphecidae and Crabronidae, notably the genera *Sceliphron*, *Chalybion*, *Trigonopsis*, and some members of *Trypoxylon*.

Mullerian mimicry Shared physical and/or behavioral attributes between related and unrelated species that are toxic or well-defended.

oviposit Act of laying eggs by a female insect.

ovipositor Organ of a female insect used to deposit eggs. This may be telescoping segments of the abdomen, or a blade-like, spear-like, or rod-like appendage protruding from the tip of the abdomen.

parasite Organism that requires another organism as a host, and feeds and grows without killing the host.

parasitoid A parasite that invariably kills its host.

parthenogenesis Production of viable offspring by a female animal without fertilization by a male.

petiole Stalk-like second abdominal segment connecting the gaster to the propodeum. *See* metasoma.

phoretic copulation Sexual intercourse in some wasps whereby the larger, winged male flies off with the smaller, wingless female. Common behavior in Thynnidae and Mutillidae.

phoretic oviposition A condition in which females of some egg parasitoids ride on the adult female host, then disembark when the host lays her eggs.

polydnaviruses Family of viruses peculiar to some wasps in the Braconidae and Ichneumonidae that exert control over the host's immune system and/or nervous system for the benefit of the growth and development of the parasitoid's offspring.

progressive provisioning Behavior in which the parent organism brings food to its offspring as needed. In sand wasps (tribe Bembicini) and some aphid wasps (*Pemphredon* spp.), the female brings prey to her larval offspring sequentially.

propodeum First segment of the abdomen in wasps, which in the Apocrita is fused to the rear of the thorax.

provisions Food supplied to her offspring by a female solitary stinging parasitoid. Typically, paralyzed insects or spiders, cached all at once (mass provisioning).

queen Reproductive female member of a colony of social insects. Gyne. A queen may be physically larger than other females, but always at the top of the dominance hierarchy.

sexual dimorphism Graphic physical differences between sexes, beyond obvious differences in genitalia. These differences include size, absence of wings, modifications of eyes or antennae.

solitary wasp Any wasp species in which females do not cooperate to rear young.

sp. Abbreviation of species, singular, after a genus name when species is unknown.

species Most specific taxonomic unit of classification. Definitions of "species" are changing, but the historically defining character is that of a population that is unable to successfully breed with another, due to incompatibility of genitalia, geographic isolation, or other factors.

spheksophobia Severe, irrational, or debilitating fear of wasps.

spp. Abbreviation of "species," plural, after a genus name, when referring to several or all species in that genus.

sting Ovipositor of many wasp species that has evolved to deliver venom.

stinging parasitoid All solitary wasp species that use their sting primarily to paralyze a host.

subfamily A subset of the family level of taxonomy that represents a collection of tribes, each of which is itself a collection of genera.

symbiosis Relationship between two species that may benefit both (mutualism), benefits one but harms the other (parasitism), or benefits one without harming the other (commensalism).

Symphyta Informal name for sawflies, horntails, and their kin. No longer a recognized taxonomic category.

thorax Middle body section of an insect between head and abdomen, to which legs and wings are attached. Internally, mostly muscle. *See* mesosoma.

trap nest Artificial nest for solitary bees and/or solitary, stinging parasitoid wasps to provide housing. A "bee condo," "bee block," or "bee hotel."

trophallaxis Exchange of food between adult insects, or between adults and larvae. A major form of communication in colonies of social insects.

venom Any chemical substance which when injected by one organism, destroys, weakens, or otherwise compromises the physiology of the recipient, usually a host or prey.

wasp A species in the order Hymenoptera that is not a bee or ant.

worker Female social wasp that does not reproduce, either through physiology (undeveloped ovaries), subordinate hierarchy, or both.

yellowjacket A social wasp in the genus *Vespula* or *Dolichovespula*, family Vespidae.

Useful References

BOOKS

Bohart, R.M. and A.S. Menke. 1976. *Sphecid Wasps of the World: A Generic Revision*. Berkeley: University of California Press. 695 pp.

Compton, John. 1955. *The Hunting Wasp*. Boston: Houghton Mifflin Co. 240 pp.

Eiseman, Charley and Noah Charney. 2010. *Tracks & Sign of Insects and Other Invertebrates: A Guide to North American Species*. Mechanicsburg, Pennsylvania: Stackpole Books. 582 pp.

Evans, Howard E. and Mary Jane West Eberhard. 1970. *The Wasps*. Ann Arbor: The University of Michigan Press. 265 pp.

Evans, Howard E. 1963. *Wasp Farm*. Garden City, New York: Natural History Press. 178 pp.

Gess, Sarah K. 1996. *The Pollen Wasps*. Cambridge, Massachusetts: Harvard University Press. 340 pp.

Goulet, Henri and John T. Huber, eds. 1993. *Hymenoptera of the World: An identification guide to families*. Ottawa: Agriculture Canada. 668 pp.

Grissell, Eric. 2010. *Bees, Wasps, and Ants: The Indispensable Role of Hymenoptera in Gardens*. Portland: Timber Press. 335 pp.

Jenner, Ronald and Eivind Undheim. 2017. *Venom: The Secrets of Nature's Deadliest Weapon*. Washington, DC: Smithsonian Books. 208 pp.

Jones, Richard. 2019. *Wasp*. London: Reaktion Books, Ltd. 207 pp.

Krombein, Karl V. 1967. *Trap-Nesting Wasps and Bees: Life Histories, Nests, and Associates*. Washington, DC: Smithsonian Press. 570 pp.

O'Neill, Kevin M. 2001. *Solitary Wasps: Behavior and Natural History*. Ithaca: Cornell University Press. 406 pp.

Preston-Mafham, Rod and Ken. 1993. *The Encyclopedia of Land Invertebrate Behavior*. Cambridge, Massachusetts: The MIT Press. 320 pp.

Rau, Phil and Nellie. 1918. *Wasp Studies Afield*. New Jersey: Princeton University Press. 372 pp.

Reinhard, Edward G. 1929. *The Witchery of Wasps*. New York: The Century Co. 291 pp.

Schmidt, Justin O. 2016. *The Sting of the Wild*. Baltimore: Johns Hopkins University Press. 257 pp.

Tinbergen, Niko. 1958. *Curious Naturalists*. Garden City, New York: Doubleday & Company, Inc. 301 pp.

Zahradnik, J. 1991. *Bees, Wasps and Ants*. London: The Hamlyn Publishing Group, Ltd. 192 pp.

ONLINE RESOURCES

Chrysis.net: **WWW.CHRYSIS.NET/** All about cuckoo wasps.

Universal Chalcidoidea Database: **WWW.NHM.AC.UK/OUR-SCIENCE/DATA/CHALCIDOIDS/DATABASE/** Global treatment of micro-Hymenoptera.

WaspWeb: **WWW.WASPWEB.ORG/CLASSIFICATION/INDEX.HTM** Classification of Afro-Tropical Hymenoptera.

BugGuide: **HTTPS://BUGGUIDE.NET/NODE/VIEW/59** Arthropods of North America north of Mexico.

iNaturalist: **WWW.INATURALIST.ORG/** Reciprocal website and database that anyone can contribute to, with projects users can participate in.

Project Noah: **WWW.PROJECTNOAH.ORG/** Similar website to iNaturlist, somewhat friendlier to younger, novice naturalists.

Evanoidea Online: **HTTP://EVANIOIDEA.INFO/PUBLIC/SITE/EVANIOIDEA/HOME** Catalog of information about evanioid wasps (ensign wasps, wild carrot wasps, and so on).

BWARS: **WWW.BWARS.COM/HOME** Bees, Wasps, and Ants Recording Society for Britain and Ireland.

Figweb: **WWW.FIGWEB.ORG/FIGS_AND_FIG_WASPS/INDEX.HTM** Comprehensive information about figs, fig wasps, and other organisms associated with figs.

Xerces Society: **HTTP://XERCES.ORG/** International invertebrate conservation organization with extensive resources and publications on pollinating insects, and so on.

Common Bees & Wasps of Ohio Field Guide: **HTTPS://OHIODNR.GOV/WPS/PORTAL/GOV/ODNR-CORE/DOCUMENTS/WILDLIFE-DOCUMENTS/BEES-WASPS-OHIO** PDF of pocket guide to wasps and bees found in Ohio, U.S.A., written by the author of this book.

Brisbane Insects: **WWW.BRISBANEINSECTS.COM/BRISBANE_WASPS/INDEX.HTML** Good overview of some common Australian wasps.

Bug Eric blog: **HTTP://BUGERIC.BLOGSPOT.COM/SEARCH/LABEL/WASPS** The author's blog, featuring many posts about wasps.

Golden Goddess
A female great golden digger, *Sphex ichneumoneus*, drinks nectar from a globe thistle (*Echinops* sp.) in a Toronto, Ontario, Canada garden. Plant native flowers for bees and you will also attract beneficial wasps.

Index

Picture Credits

All illustrations by Sandra Pond.

Photographs: p3: Courtesy of Life on White/Alamy Stock Photo. p4 & 5: Courtesy of alslutsky/Shutterstock (far left, left, right & far right); courtesy of Cornel Constantin/Shutterstock (center). p7: Courtesy of Tom Barnes. p8: Courtesy of Danny Radius/Shutterstock. p9: Courtesy of Alfred Daniel. p11: Courtesy of Daniel Prudek/Shutterstock. p14: Courtesy of Filipao Photography/Shutterstock. p15: Courtesy of Bjoern Wylezich/Shutterstock. p16: Courtesy of Kent and Donna Dannen/Science Photo Library. p17: Courtesy of Bjoern Wylezich/Shutterstock (top); courtesy of John Cancalosi/Nature Picture Library (bottom). p18: Courtesy of Mr. Background/Shutterstock. p21: Courtesy of Colin Marshall/agefotostock/Alamy Stock Photo. p25: Courtesy of Gerhard Koertner/Avalon/Photoshot License/Alamy Stock Photo. p27: Courtesy of irin-k/Shutterstock (top); courtesy of Cristina Romero Palma/Shutterstock (bottom). p28: Courtesy of Denis Vesely/Shutterstock. p29: Courtesy of Cosmin Manci/Shutterstock. p30: Courtesy of Malcolm Park sciences/Alamy Stock Photo. p31: Courtesy of Michael Caterino (top); courtesy of alslutsky/Shutterstock (center top); courtesy of Pierre Bornand (center bottom & bottom). p33: Courtesy of Danny Radius/Shutterstock (top & bottom). p34: Courtesy of Cherdchai Chaivimol/Shutterstock. p35: Courtesy of Gerhard Koertner/Avalon/Photoshot License/Alamy Stock Photo. p36: Courtesy of alslutsky/Shutterstock. p37: Courtesy of Danny Radius/Shutterstock (top); courtesy of Dennis Kunkel Microscopy/Science Photo Library (bottom). p38: Courtesy of Stefan Rotter/Shutterstock. p39: Courtesy of blickwinkel/Lenke/Alamy Stock Photo (bottom left); courtesy of photowind/Shutterstock (top right). p41: Courtesy of Biehler Michael/Shutterstock. p42: Courtesy of alslutsky/Shutterstock. p44: Courtesy of Tongaaa/Shutterstock. p45: Courtesy of Milan Radisics/Nature Picture Library (top); courtesy of Eric Stavale (center). p46: Courtesy of Kristi Ellingsen. p47: Courtesy of Pierre Bornand. p49: Courtesy of Sean McVey/Shutterstock. p51: Courtesy of Wanchat M/Shutterstock (top); courtesy of Tomasz Klejdysz/Shutterstock (bottom). p52: Courtesy of J.J. Gouin/Shutterstock. p53: Courtesy of Jeremy A. Casado/Shutterstock (top & bottom). p54: Courtesy of Maciej Olszewski/Alamy Stock Photo. p55: Courtesy of Mario Saccomano/Shutterstock (top); courtesy of Sarah2/Shutterstock. p56: Courtesy of blickwinkel/G. Kunz/Alamy Stock Photo. p57: Courtesy of Elmanther Lee/Shutterstock. p58: Courtesy of Sean McVey/Shutterstock. p59: Courtesy of Daniel Knop/Shutterstock (all images). p60: Courtesy of Jay Ondreicka/Shutterstock. p61: Courtesy of Sloan Tomlinson (top); courtesy of Brett Hondow/Shutterstock (center); courtesy of Stephen Bonk/Shutterstock (bottom). p63: Courtesy of Jay Peter Yeeles/Shutterstock (top); courtesy of Pascal Guay/Shutterstock (bottom). p64: Courtesy of Possent phsycography/Shutterstock. p65: Courtesy of Emanuele Biggi/FLPA/Nature in Stock (top, second from top and bottom); courtesy of FLPA/Alamy Stock Photo (third image from top). p67: Courtesy of Nigel Eve/Shutterstock (top); courtesy of Kazakov Maksim/Shutterstock (bottom). p69: Courtesy of Cornel Constantin/Shutterstock. p71: Courtesy of Javier Torrent, VW Pics/Science Photo Library. p72: Courtesy of Cornel Constantin/Shutterstock. p73: Courtesy of alslutsky/Shutterstock (top); courtesy of yod67/Shutterstock (bottom). p75: Courtesy of alslutsky/Shutterstock (top); courtesy of Danny Radius/Shutterstock (bottom). p77: Courtesy of UbjsP/Shutterstock. p78: Courtesy of alslutsky/Shutterstock (top right); courtesy of Henrik Larsson/Shutterstock (bottom left); courtesy of alslutsky/Shutterstock (bottom right). p79: Courtesy of alslutsky/Shutterstock (all images). p80: Courtesy of Sari ONealShutterstock (bottom). p82: Courtesy of Steve Heap/Shutterstock. p83: Courtesy of Anton Kozyrev/Shutterstock. p85: Courtesy of Eric Isselee/Shutterstock. p87: Courtesy of Igor Kovalenko/Shutterstock. p88: Courtesy of Lenti Hill/Shutterstock. p90: Courtesy of Natasha Mhatre. p92: Courtesy of thatmacroguy/Shutterstock. p93: Courtesy of John Ceulemans/Shutterstock (top); courtesy of J.M. Abarca/Shutterstock (bottom). p94: Courtesy of Melvyn Yeo/Biosphoto/Alamy Stock Photo. p96: Courtesy of Protasov AN/Shutterstock. p97: Courtesy of Tomasz Klejdysz/Shutterstock (bottom). p98: Courtesy of Nigel Cattlin/Alamy Stock Photo. p100: Courtesy of Natasha Mhatre. p101: Courtesy of Natasha Mhatre (top); courtesy of Ruth Swan/Shutterstock (bottom). p102: Courtesy of phichak/Shutterstock. p103: Courtesy of Protasov AN/Shutterstock. p105: Courtesy of Daniel Prudek/Shutterstock. p109: Courtesy of Nikola Rahme/Shutterstock (bottom). p111: Courtesy of Alex Wild (top and middle); courtesy of Stoil Ivanov (bottom). p112: Courtesy of Jena Johnson. p113: Courtesy of Nigel Eve/Shutterstock (top); courtesy of AlessandroZocc/Shutterstock (bottom). p116: Courtesy of AJCespedes/Shutterstock. p117: Courtesy of Jena Johnson (top). p119: Courtesy of Eric R. Eaton. p119: Courtesy of Cindy Bingham Keiser (all images). p121: Courtesy of enterlinedesign/Shutterstock (middle); courtesy of R K Hill/Shutterstock (bottom). p122: Courtesy of Ryan Silva/Shutterstock. p124 & 125: Courtesy of Jena Johnson. p127: Courtesy of John Abbott/Nature Picture Library/Alamy Stock Photo. p128: Courtesy of Elya Vatel/Shutterstock. p129: Courtesy of Stefan Rotter/Shutterstock. p130 & 131: Courtesy of Premaphotos/Alamy Stock Photo. p133: Courtesy of Revilo Lessen/Shutterstock. p135: Courtesy of Danny Radius/Shutterstock. p136: Courtesy of BrunoGarridoMacias/Shutterstock. p137: Courtesy of Nigel Cattlin/FLPA/Nature in Stock (top); courtesy of Natasha Mhatre (bottom). p139: Courtesy of Lubos Chlubny/Shutterstock (top); courtesy of Cathy Withers-Clarke/Shutterstock (bottom). p141: Courtesy of kurt_G/Shutterstock (top); courtesy of yod67/Shutterstock (bottom). p142: Courtesy of Denis Crawford/Alamy Stock Photo. p143: Courtesy of vblinov/Shutterstock (top); courtesy of Agung Y.P/Shutterstock (bottom). p145: Courtesy of Shivkumar Sharma/Shutterstock. p147: Courtesy of Tom Barnes. p148: Courtesy of Jeff Gruber. p149: Courtesy of Clarence Holmes Wildlife/Alamy Stock Photo (top); courtesy of Conan P/Shutterstock (middle); courtesy of Eduardo Dzophoto/Shutterstock (bottom). p150: Courtesy of David Hughes. p151: Courtesy of Ctatiana/Shutterstock. p152: Courtesy of Sriyana/Shutterstock. p153: Courtesy of Danny Radius/Shutterstock (top); courtesy of Clifford Pugliese/Shutterstock (bottom). p154: Courtesy of Mart Smit/Nature in Stock (top). p154 & 155: Courtesy of David Hughes (bottom). p155: Courtesy of Seanoneillphoto/Shutterstock (top left). p157: Courtesy of Vinicius R. Souza/Shutterstock (top left); courtesy of Jeff Gruber (right); courtesy of Simon Shim/Shutterstock (bottom left). p159: Courtesy of stockfoto/Shutterstock. p161: Courtesy of Jeff Gruber. p162: Courtesy of Drosophila Photograph/Shutterstock. p163: Courtesy of Jeff Gruber (top & middle); courtesy of stockfoto/Shutterstock (bottom). p164: Courtesy of D. Kucharski K. Kucharska/Shutterstock. p165: Courtesy of Jeff Gruber (bottom). p166: Courtesy of Vinicius R. Souza/Shutterstock (top); courtesy of Premaphotos/Alamy Stock Photo (bottom). p167: Courtesy of zairiazmal/Shutterstock. p168: Courtesy of Matthew Bertone. p170: Courtesy of Frank Deschandol/© Frank Canon. p171: Courtesy of Premaphotos/Alamy Stock Photo. p172: Courtesy of Ken Griffiths/Shutterstock. p173: Courtesy of pixelworlds/Shutterstock. p174: Courtesy of Jeff Gruber. p175: Courtesy of Protasov AN/Shutterstock (top right & bottom left). p176: Courtesy of Jeff Gruber. p177: Courtesy of Melvyn Yeo/Biosphoto/Alamy Stock Photo. p179: Courtesy of Zety Akhzar/Shutterstock. p181: Courtesy of Kuttelvaserova Stuchelova/Shutterstock. p182: Courtesy of Super Prin/Shutterstock. p183: Courtesy of Neil Bowman/FLPA/Nature in Stock (top); courtesy of Omariam/Shutterstock (bottom). p184: Courtesy of Christiane Godin/Shutterstock. p185: Courtesy of Vinicius R. Souza/Shutterstock (top left); courtesy of Jeff Gruber (top right); courtesy of kezza/Shutterstock (bottom). p186: Courtesy of Melinda Fawver/Shutterstock. p187: Courtesy of Ross Piper. p189: Courtesy of Andrew Polaszek/Natural History Museum/Wellcome Collection (CC by 4.0). p190: Courtesy of pragari/Shutterstock (top); courtesy of Blue bee/Shutterstock (bottom). p191: Courtesy of I Wayan Sumatika/Shutterstock. p192: Courtesy of thatmacroguy/Shutterstock. p193: Courtesy of Vinicius R. Souza/Shutterstock. p197: Courtesy of Daniel Prudek/Shutterstock. p199: Courtesy of Jeroen Mikkers/Shutterstock. p201: Courtesy of Brais Seara/Shutterstock (top left); courtesy of Elena Lebedeva-Hooft/Shutterstock (top right). p202: Courtesy of Gorg Evd/Shutterstock. p203: Courtesy of Geza Farkas/Shutterstock (top); courtesy of 1981 Rustic Studio kan/Shutterstock (bottom). p204: Courtesy of icosha/Shutterstock. p205: Courtesy of Warut Prathaksithorn/Shutterstock (top); courtesy of Chronicle/Alamy Stock Photo (bottom). p206: Courtesy of Science History Images/Alamy Stock Photo. p207: Courtesy of Natalia Price-Cabrera/Chris Gatcum (top); courtesy of Devon Henderson (bottom). p208: Courtesy of Mullwell/Shutterstock. p209: Courtesy of Bildagentur Zoonar GmbH/Shutterstock (top left); courtesy of Angela Rohde/Shutterstock (top right); courtesy of Daniel Prudek/Shutterstock (bottom). p210: Courtesy of NASA; p211: Courtesy of Nigel Cattlin/FLPA/Nature in Stock (top left); courtesy of Erik Agar/Shutterstock (right). p212: Courtesy of David Hughes. p213: Courtesy of Okqy Purnama Setiawan/Shutterstock (bottom left); courtesy of Melinda Fawver/Shutterstock (top right). p215: Courtesy of Jena Johnson. p216, 217, 218 & 219 (top): Courtesy of Matthew Bertone. p219: Courtesy of alslutsky/Shutterstock (bottom). p220: Courtesy of alslutsky/Shutterstock (top); courtesy of Geraldine Buckley/Alamy Stock Photo (bottom). p221: Courtesy of Tom Barnes (top left); courtesy of Matthew Bertone (top right); courtesy of Henri Koskinen/Shutterstock (bottom). p222: Courtesy of Matthew Bertone (top); courtesy of DeRebus/Shutterstock (bottom). p223: Courtesy of Eric R. Eaton (top); courtesy of Matteo Omied/Alamy Stock Photo (bottom). p224: Courtesy of Kristi Ellingsen. p225: Courtesy of fendercapture/Shutterstock (top); courtesy of MM-Fotos/Shutterstock (bottom). p226: Courtesy of Brian Valentine (top); courtesy of Sandra M. Austin/Shutterstock (bottom). p227: Courtesy of Brian Valentine. p228: Courtesy of Pierre Bornand (top); courtesy of Jena Johnson (bottom). p229: Courtesy of Pierre Bornand (top); courtesy of Tim Leppek (bottom). p230: Courtesy of Natasha Mhatre (top); courtesy of Ihor Hvozdetskyi/Shutterstock (bottom). p231: Courtesy of Matthew Bertone. p232: Courtesy of ChinKC/Shutterstock (top); courtesy of Matthew Bertone (bottom). p233: Courtesy of khlungcenter/Shutterstock (top); courtesy of Jean Lecomte/Biosphoto/Alamy Stock Photo (bottom). p234: Courtesy of Frode Jacobsen/Shutterstock (top); courtesy of Alfred Daniel (bottom). p235: Courtesy of Jena Johnson (top); courtesy of Pierre Bornand (bottom). p236: Courtesy of Pierre Bornand (top); courtesy of Denis Crawford/Alamy Stock Photo (bottom). p237: Courtesy of Roberto Michel/Shutterstock (top); courtesy of Matthew Bertone (bottom). p238: Courtesy of Tom Barnes (top); courtesy of Matthew Bertone (bottom). p239: Courtesy of Pierre Bornand (top); courtesy of Jude Black/Shutterstock (bottom). p240: Courtesy of yod67/Shutterstock (top); courtesy of Rifki_stn/Shutterstock (bottom). p241: Courtesy of Matthew Bertone (top); courtesy of Pujiyono/Shutterstock (bottom). p242: Courtesy of kurt_G/Shutterstock (bottom). p243: Courtesy of Bryan Reynolds/Alamy Stock Photo (top); courtesy of alslutsky/Shutterstock (bottom). p244: Courtesy of Ranjit_Mahara_Photography/Shutterstock (top); courtesy of Matthew Bertone (bottom). p245: Courtesy of Eileen Kumpf/Shutterstock (top); courtesy of Pierre Bornand (bottom). p247: Courtesy of Alexander Chernyy/Shutterstock. p251: Courtesy of Paul Reeves Photography/Shutterstock. p256: Courtesy of Matteo Omied/Alamy Stock Photo.

Acknowledgments

I could not have undertaken this book project without the invitation of Kate Shanahan of UniPress Books. Kate put me in the most professional hands of Natalia Price-Cabrera, my project manager, Nigel Browning for all things business-related, and Robert Kirk of our literary partner, Princeton University Press. Without the stunning artwork of illustrator Sandra Pond, this volume would not be nearly as captivating, comprehensible, and aesthetically beautiful. Thanks too to Sandra Zellmer for the elegant book design. We also owe great thanks to Tom Barnes, Matthew Bertone, Cindy Bingham Keiser, Pierre Bornand, Alfred Daniel, Frank Deschandol, Kristi Ellingsen, Jeff Gruber, Devon Henderson, Laura and David Hughes, Stoil Ivanov, Jena Johnson, Tim Leppek, Natasha Mhatre, Ross Piper, Eric Stavale, Sloan Tomlinson, Brian Valentine, Alex Wild and the many other photographers who contributed images via stock agencies. I assume sole responsibility for errors of fact or omission, spelling, and grammar.

I would be remiss in failing to acknowledge the lifelong impact of mentors past and present in shaping my fascination with wasps. The late George R. Ferguson at Oregon State University was instrumental in his patient guidance and support. Howard E. Evans was a kind and generous correspondent. I have been privileged to spend many a fine day in the field with Justin O. Schmidt, no stings required. Justin's inquisitive mind and ability to formulate rigorous tests for his hypotheses continues to inspire. Edward "Eric" Grissell remains a mentor in both entomology and writing. Critical review of the chapter on wasp evolution was provided by world authorities: James M. Carpenter, American Museum of Natural History; Michael Sharkey, University of Kentucky; Susanne Schulmeister; George O. Poinar, Jr., Oregon State University; Doug Yanega, University of California, Riverside. Special thanks to MaLisa Spring and Gwen Pearson for reminding me that Sci-hub is the greatest thing on the internet, ever.

Most importantly, I thank my wife, Heidi, my lifeline for all endeavors, enduring all the angst and stress that comes with projects like this, and making sure I stay fortified with excellent food, and mentally healthy with regular hikes and excursions. She also fixes and updates all things tech and electronic in our household, as these devices have advanced far beyond my comprehension. It is my honor to be her spouse.